Children of the Stars

Are we alone in the Universe? What is our place in it? How did we get here? We have long searched for the answers to questions such as these, and scientists are beginning to find some of the answers. In this book, Daniel Altschuler provides the reader with the elements to understand the questions and their answers as far as we know them. He explores subjects from physics and astronomy to geology and palaeontology. Along the way he touches on topics of great popular appeal such as the search for life on other worlds and the hazards of asteroid impacts. The author writes in an engaging and readable style with wit, warmth and erudition at a level that any interested reader can understand.

DANIEL R. ALTSCHULER, an experienced researcher, educator, and science administrator, is director of the National Astronomy and Ionosphere Center's Arecibo Observatory and a member of the faculty of the Physics Department of the University of Puerto Rico, at Rio Piedras.

The Arecibo Observatory is the site of the largest telescope on Earth. Daniel, who was born in Montevideo, Uruguay, the son of German immigrants, completed an Engineering degree from Duke University, and obtained his PhD in physics from Brandeis University. He has given numerous public lectures in many places from Uruguay to South Africa, and has appeared in radio and television shows including *Good Morning America* and *The Learning Channel*. He places the construction of a Visitor Center at the Arecibo Observatory, visited by over 120 000 persons every year, among his most satisfying achievements. The other is scoring a decisive goal for the team of the Max Planck Institut für Radioastronomie, in Germany, where he spent two years as a visiting scientist. Dr. Altschuler is a member of the American Astronomical Society and of the International Astronomical Union. He states "I have always been concerned about the poor understanding of scientific topics by a large segment of the population. It is not enough for scientists alone to understand the workings of nature. It is important that every citizen understands what scientists have been able to learn, not only because it is interesting, truly fascinating, but also because difficult decisions must be made by all, and can only be made with a clear understanding of the issues . . . a good part of this state of affairs has been the consequence of the little interest and less time taken by scientists to communicate with the public. My book is an effort to remedy this situation."

children of the Stars

Our Origin, Evolution and Destiny

Daniel R. Altschuler

PUBLISHED BY THE PRESS SYNDICATE OF THE UNIVERSITY OF CAMBRIDGE
The Pitt Building, Trumpington Street, Cambridge, United Kingdom

CAMBRIDGE UNIVERSITY PRESS
The Edinburgh Building, Cambridge CB2 2RU, UK
40 West 20th Street, New York, NY 10011–4211, USA
477 Williamstown Road, Port Melbourne, VIC 3207, Australia
Ruiz de Alarcón 13, 28014 Madrid, Spain
Dock House, The Waterfront, Cape Town 8001, South Africa

http://www.cambridge.org

First published 2002

Printed in the United Kingdom at the University Press, Cambridge

Typeface Swift 9.5/13.5 pt   *System* QuarkXPress™   [SE]

A catalogue record for this book is available from the British Library

Library of Congress Cataloguing in Publication data
Altschuler, Daniel R.
    Children of the stars : our origin, evolution, and destiny / Daniel R. Altschuler.
        p.   cm.
    Includes bibliographical references and index.
    ISBN 0 521 81212 7
        1. Astronomy–Popular works. I. Title.
    QB44.3 .A58 2002
    520–dc21   2001043172

ISBN 0 521 81212 7 hardback

Contents

Some say the world will end in fire,
Some say ice.
From what I've tasted of desire
I hold with those who favor fire.
But if it had to perish twice,
I think I know enough of hate
To say that for destruction ice
Is also great
And would suffice.

Robert Frost (1874–1963)

Prelude

Look at your hand. It is made of atoms, mostly atoms of carbon, oxygen, and hydrogen. These atoms did not always exist; they were produced inside stars.

Look at the ring on your finger. It is also made of atoms, maybe of gold or platinum. These were created during the death paroxysm of a massive star, a supernova explosion 5 billion years ago. Just like us, stars are born, live mostly a quiet life and die.

Open your hand toward the Sun and feel its heat. It is the energy of life.

Look at your hand again. It has five fingers, as do the hands of some other animals. This is not a coincidence, but points to the profound fraternity of all life forms on Mother Earth, one of millions of planets orbiting other distant stars. Perhaps we are not alone in this vast Universe, we would dearly like to know.

This book tells the story of the incredible events that, starting with the stars, lead to us. It will change forever both the way you see yourself and the way you see our world.

And when you look at the world do not be deceived by how large and robust it seems to be. The place we live in, the biosphere, is a delicate veneer on the surface of this tiny dot orbiting the Sun. We must treat it with respect since it is fragile, and not abuse its resources since they are limited.

> The aim of science is not to open the door to infinite wisdom, but to set a limit to infinite error
> *Bertold Brecht in* Life of Galileo *(1938)*

When the Moon places itself directly between the Sun and the Earth, it blocks out the Sun's light, producing a solar eclipse. In this photograph obtained by astronomer Tunç Tezel from Turkey, the day before the solar eclipse of August 11, 1999, the Moon, illuminated almost entirely from the back by the Sun (which is out of the picture to the lower left), shows up as a thin sliver of light. Mercury, the closest planet to the Sun, is never far from it in the sky and therefore normally difficult to see. On this photo the small planet is clearly visible at the bottom right. The dark side of the Moon is dimly illuminated by the light from an almost "full Earth." (Tunç Tezel)

Preface

Should you have time one fine autumn day, take the afternoon off and drive to a location in the countryside, away from the lights and sounds of the city, from where you can see the night sky in all its calm beauty. Find a nice spot, and wait until the Sun sets and your side of our planet faces the majesty of the dark sky. You are now looking at the Universe.

At twilight you might see a thin curved sliver of light, and with some care you will notice that it is the bright rim on the edge of a dark disk. It is our Moon, a wonderful sight, and when it is almost new a thin crescent of light is visible. The dark part is just about visible because it is dimly illuminated by sunlight which is reflected from the Earth. If you wait long enough, you will probably also see a "shooting star," not a star at all, but the incandescent trail left by a small particle – meteoritic dust – which by chance has entered our atmosphere at high speed and, heated up by friction, produces light. This meteoroid, if large enough, might survive its fiery trip and hit the surface of the Earth to be found as a meteorite. Sometimes it might even be large enough to make a crater of considerable dimensions, and on very, very rare occasions it could be so large that its impact would have catastrophic consequences.

Now lie on your back and spend some time looking at the awesome display painted by a myriad of twinkling stars which have quietly appeared against the black sky. You may also be able to recognize the occasional planet. After a while, as your eyes adjust to darkness, you will realize that a band of diffuse light crosses the sky from horizon to horizon. If you look through binoculars, you will find that this light is composed of countless stars, as was first discovered by Galileo Galilei about 400 years ago, when in 1609 he first used a small telescope to look at the night sky. This is the "Galaxias" of the Greeks, the "Via Lactea" of the Romans, the Milky Way to us, which today we know to be our galaxy, our home in this immense Universe. In antiquity, and for many cultures around the world, the Milky Way was thought of as either a river of heaven, a great silvery serpent, or a path in the sky. It is a gigantic flattened disk-like system composed of billions of stars which, when viewed from our position within it, appears to be a band of light across the sky.

After a while you will experience a wondrous feeling, a nostalgic feeling which comes from very deep within, but also great joy, somehow related to your profoundest intuition, telling you that you are looking at our origin. *Our origin is in the stars.*

This astonishing discovery is the result of the efforts of scientists who over the last few centuries have struggled to uncover the secrets of nature. This is the goal of science: to understand the natural world through observation, experimentation, and computation. Science is also the foundation of the many technological developments (both good and bad) which drastically transformed life in the twentieth century. Most importantly, science provides the framework for understanding Life, the Universe, and Everything, at least as far as we have yet been able to.

I have written this book because the resulting story of the origins of matter and life, and their evolution, is an awesome story. It is better than science fiction, far better than all those fake UFO stories published in some newspapers and magazines, and even better than those great movies about such topics. This is because of one important and particular reason: *this* story is real. Without doubt, the unraveling of this story is the crowning intellectual achievement of the second millennium AD, and something that every one of us should be proud of.

I hope that this book will let you appreciate life in a new way. We are so preoccupied by our daily lives that we forget that we are only here for a very short visit. And then one day we are dead. I hope that after reading this book you will see your fellow human beings and the world in a different light. This might allow you to think in new ways which might just help extricate us from the frightening environmental predicament into which we have unwittingly placed ourselves.

This story is the culmination of a long development toward a new view of the Universe, a view which proposes a historical process. In spite of the opposition from some who believe that all there is to know about our Universe is written in ancient texts, this is an evolutionary view in the grandest sense, way beyond anything that Charles Darwin could have imagined (although the evolution of the Universe is not the same as Darwinian evolution). It involves the entire Universe which, over a very long time, created matter until it evolved to the point where planets, such as Earth, could be formed and become fertile for life to arise.

This is a very different outlook from the comfortable position held until about 500 years ago, which asserted that everything was as it always had been, since God's creation. This creation was said to have happened not very long ago, but everything would continue to be exactly as it was then, forever. The Universe was perceived as static and was therefore a safe place. Over the past 500 years, those who proposed the new evolutionary view clashed with two opponents. One has been represented by the healthy resistance provided by the scientific establishment, and the other by religious or other dogma, a stubborn need to stick to literal scripture no matter what the evidence might be, as was the case with Aristotelian physics which was dogmatically adhered to for over 1000 years. There were learned people who refused to look through Galileo's telescope, lest this might shatter their convictions. Scientific conservatism is healthy because it demands proof in support of all new ideas, proof based on repeatable measurements which are sometimes difficult to produce. This is what gives science its power and fecundity. On the other hand, dogma is sterile, locking into a dark chamber the one quality which distinguishes us from other animals: our minds.

Contrary to some popular perceptions, science is not a cold matter-of-fact endeavor ruled only by unquestionable facts. Facts can and should be questioned, and must be interpreted in the light of some framework or theory. In the end, after careful examination, the facts triumph. To a certain extent, which we try to minimize, science is also influenced by personal beliefs, cultural backgrounds, and social climate. This is because scientists *are* human. However, new ideas, no matter how disagreeable or difficult to come to terms with, will eventually be accepted in the light of the evidence. Such evidence has led to the acceptance of a Sun-centered planetary system, continental drift, and the origin of species. Each of these created quite a storm, a hurricane that made the ship of science change its course. On the other hand, dogma of every kind will not be so understanding of new ideas if they are perceived as running counter to it. The ship of dogmatism will surely run aground, or capsize and sink, at some point in the voyage. We have witnessed this throughout history.

This story is the result of the courage and determination of individuals who have dedicated their lives to the study of nature, sometimes at great personal risk and sacrifice. I have sprinkled the story with a few brief vignettes of these heroes, really not enough to do them justice, but I hope enough to pay them homage. Of course, these heroes – Copernicus, Darwin, Wegener, to name just a few – only represent the most prominent of those who participated in the process of discovery, which was not as clear and straightforward as you might think. There were false starts, intermediate steps, and excursions toward dead ends, on the way to the breakthroughs that led to a better understanding of our world. These were men and women living in difficult times, working with the limited tools available one hundred or more years ago, writing by candle light, and traveling by sailing ship or horse-drawn carriage. They often depended for their livelihood on the erratic support of some wealthy patron. It is difficult to imagine how they achieved what they did, and to understand what drove and inspired them to sometimes risk their lives in defense of their convictions.

Today we live in a world populated by sophisticated instruments. Giant particle accelerators probe the smallest dimensions of matter, machines analyze the molecules of life, large telescopes, both on Earth and in space, look out to the farthest reaches of the Universe, and spacecraft scrutinize the Earth and other planets of the solar system. Many nations have established agencies such as the US National Science Foundation (NSF), the National Institutes of Health (NIH), and the National Aeronautics and Space Administration (NASA), to support the efforts of scientists who gather an ever-increasing flow of data which become the basis for new knowledge. This is not a luxury, catering only to the aspirations of the academic community, but a necessity for our progress and our survival.

Our power over nature has given us the ability to destroy our fragile biosphere. Only a very deep and precise knowledge of the world around us will allow us to understand the consequences of our actions, and so avoid the dire effects which are impacting life on a global scale. In this book you will learn that we are so intimately connected with nature that any thought of independence can only lead to disaster. However, it is not enough for scientists alone to understand nature. It is also important that every citizen understands what scientists have been able to learn, not only because it is interesting, truly fascinating, but also because difficult decisions must be made by all, and can only be made with a clear understanding of the issues. In a world which is very dependent on technology, participatory democracy can only work at its best with a scientifically literate population. How else are we going to be able to judge the value of sending a probe to explore Mars, the need to reduce our emissions of carbon dioxide, the use of nuclear bombs to protect us from asteroids, or the effects of introducing genetically modified foods? How else will we be able to distinguish between esoteric gobbledygook that speaks of "metaphysical vortices of rotational energy" and real natural phenomena? Furthermore, at a more essential level, science reveals for us the profound and extraordinary beauty of nature.

I have simplified many aspects of the story, not discussing the complex details which might be of greatest interest to a scientist. Some aspects of the story are only poorly understood today while others are still controversial and could be modified by future research. This is the nature of science. This is *not* a textbook and I have not tried to develop each topic systematically and completely, considering all the alternatives and putting in all the "perhaps" and all the "maybes". To do so would require each chapter to be converted into a 200-page text written by an expert, and then you would most likely not read it. I have, however, not shied away from explaining some basic ideas and including some numbers, without which this story cannot really be understood. I have told the story as if I were telling it to a good (and patient) friend, and have kept it short so as not to tax your patience. It is a somewhat eclectic story in which I have described what I find to be some of the most fascinating and important findings to come from science, findings that bear on what we are, our place in the Universe, and our future on Earth.

I have used approximate numbers for the measurements of different quantities. What is important for an understanding of the story is not the fact that the diameter of our Earth at the equator is exactly 12 713.51 km, but that 13 000 km is about $\frac{1}{30}$ of the distance from Earth to the Moon. I hope to whet your appetite to read more about the topics covered in this book, and to this end have suggested some books for further reading at the end.

We recently celebrated the arrival of a new millennium as we passed from 1999 to 2000, notwithstanding that the new millennium did not arrive until the year 2001 (because there was no year zero). It is not clear to me what all the commotion was about, except that it was a good excuse to celebrate, and we need all the excuses we can find for that. There is nothing very special about having gone 2000 times about the Sun since the time that those in the Christian world started counting. The Earth, with all its inhabitants, has done this several billion times. We travel about the Sun at the enormous speed of 67 000 miles every hour, which is about 1000 times faster than the speed you travel in your car. In the last 2000 years we have covered a distance of about 1000 billion miles going in circles, something we all seem to do frustratingly often. To remind you: 1 billion equals 1000 million. Had we gone in a straight line instead, we would now be just two-tenths of the way to the *nearest* star. Our place in the Universe, whatever it might be in philosophical terms, is clearly a very small one physically.

If after reading this book you become motivated to look at the magnificent cosmic display as I have suggested at the beginning, then my efforts in writing it will have been well worth while.

Or you might prefer to think that the answer to the question of Life, the Universe and Everything is . . . *forty-two.*

This last statement comes from one of my favorite books, *The Hitchhiker's Guide to the Galaxy* by Douglas Adams, and its various sequels. The quotations at the beginning of each chapter of this book are his.

Writing this book has taken an inordinate amount of my time over several years – long nights of study, writing and rewriting, and rewriting yet again, with sometimes a feeling of never getting to the end. I was inspired by the illusion that reading this book would make a difference. I hope it does. I hope that the next time you contemplate a red sunset, or look at a fellow human being or at another animal, or watch as a cloud drifts by in the sky, you do so with different eyes. I hope that, whatever you do in your life, you will ponder the consequences of your actions on our biosphere, even if these are minute.

The realization that the book's sales depend on a "market" which shuns books about science is not encouraging. I want to share this book with "everyone" – slightly more that the 10000 persons that might buy it if it becomes a great success (so I have been told). The thought that if I had written about the love life of aliens instead (those we study at a large hidden laboratory at the Arecibo Observatory) it might have become a bestseller disturbs me very much. But that is the real world.

Anyway, now the book is in your hands, and I hope you enjoy it and learn something from it. I welcome your comments, which you can email to: stern@ naic.edu.

Author's acknowledgments

My wife Celia not only helped me with the grammar of the text, but also taught me a lot about the more complex grammar of life. She knows how much I care for all this, and encouraged me when I felt like forgetting all about it. To her I dedicate this book.

Many persons and events in my life have influenced my way of seeing things. My father taught me that life is serious business, Eddy taught me that living is an art, and my mother taught me that life is not serious business. Emmy Link taught me to respect and love our planet, and "la barra" in Montevideo, that beautiful bunch of kids with whom I grew up, taught me the meaning of friendship. Claudio Benski inspired me.

My teachers were many, too many to mention, but I remember with affection Gregorio Treibich, and my "cosmography" teacher Conrado Schneider, who first opened my eyes to the Universe.

Chris Salter (Arecibo, Puerto Rico) helped to make a draft into something readable with his wonderful command of English and of astronomy. I thank my friends and colleagues: José Alonso (Arecibo, Puerto Rico), Fernando Diaz (San Juan, Puerto Rico), Carlo Giovanardi (Florence, Italy), Riccardo Giovanelli (Ithaca, New York), Jon Hagen (Arecibo, Puerto Rico), Margarita Irizarry (San Juan, Puerto Rico), Guillermo Irizarry (San Juan, Puerto Rico), Carmen A. Pantoja (San Juan, Puerto Rico), Giselle Petrides (Montevideo, Uruguay), Jorge Santiago (Philadelphia, USA), Matthew Windham (Adelaide,

Australia), and Kurt Ziehboldt (Hamburg, Germany), who read the manuscript and with their critique and comments helped to improve it. José F. Salgado (Chicago, USA) also prepared some of the diagrams which illustrate the text.

Paul and Paula Morgenstein (New York, USA) wrote me a letter that gave me the final nudge to embark on this arduous task. Finally, I wish to thank all those unsung heroes of the scientific enterprise: engineers, technicians, computing experts, and yes (why not?), administrators, because, without their efforts and dedication, this story could not be told.

Acknowledgments

The author and publishers would like to thank those who gave permission to reproduce copyright material.

Full acknowledgments for all illustrations reproduced in the book are given in the figure captions.

The poem "Fire and Ice" on the epigraph page (p. vi) is reproduced from *The Poetry of Robert Frost* edited by Edward Connery Lathem. Copyright 1923. © 1969 by Henry Holt and Co., Copyright 1951 by Robert Frost. Reprinted by permission of Henry Holt and Company LLC.

All the chapter opening quotes (except the one for Chapter 3) are from *Six Stories by Douglas Adams: The Ultimate Hitchhiker's Guide*, Wings Books, New York, 1996. Sequels title and page numbers for the compendium volume are given in the relevant footnotes.

The text reproduced in Appendix A (the story of the Sibylline books) is taken from D. Adams and M. Carwardine, *Last Chance to See*, 2nd edn, Pan Books Ltd and William Heinemann Ltd, London, 1991, pp. 196–9.

The text reproduced in Appendix C is the World Scientists' Warning to Humanity, which was written and spearheaded by the late Henry Kendall, former Chair of the board of directors of the Union of Concerned Scientists, 2 Brattle Square, Cambridge, MA 02238 (www.ucsusa.org).

Further acknowledgements for various text extracts are given in footnotes.

Common abbreviations

AURA   Association of Universities for Research in Astronomy
      (www.aura_astronomy.org/)
CFHT   Canada, France, Hawaii Telescope (www.cfht.hawaii.edu/)
ESA   European Space Agency (www.esa.int/export/esaCP/index.html)
ESO   European Southern Observatory (www.eso.org/)
GSFC   Goddard Space Flight Center (www.gsfc.nasa.gov/)
JPL   Jet Propulsion Laboratory (www.jpl.nasa.gov/)
LBNL   E. O. Lawrence Berkeley National Laboratory (www.lbl.gov/)
NAIC   National Astronomy and Ionosphere Center (www.naic.edu/)
NASA   National Aeronautics and Space Administration (www.nasa.gov/)
NOAA   National Oceanic and Atmospheric Administration
      (www.noaa.gov/)
NOAO   National Optical Astronomy Observatories (www.noao.edu/)
NHGRI   National Human Genome Research Institute (www.nhgri.nih.gov/)
NSF   National Science Foundation (www.nsf.gov/)
NURP   National Undersea Research Program (www.nurp.noaa.gov/)
SAAO   South Africa Astronomical Observatory (www.saao.ac.za/)
SOHO   Solar Heliospheric Observatory (www.nascom.nasa.gov/)
STScI   Space Telescope Science Institute (www.stsci.edu/)
USGS   United States Geological Survey (www.usgs.gov/)
VLT   Very Large Telescope (ESO)
WIYN   Wisconsin, Indiana, Yale, NOAO (www.noao.edu/wiyn/wiyn.html)

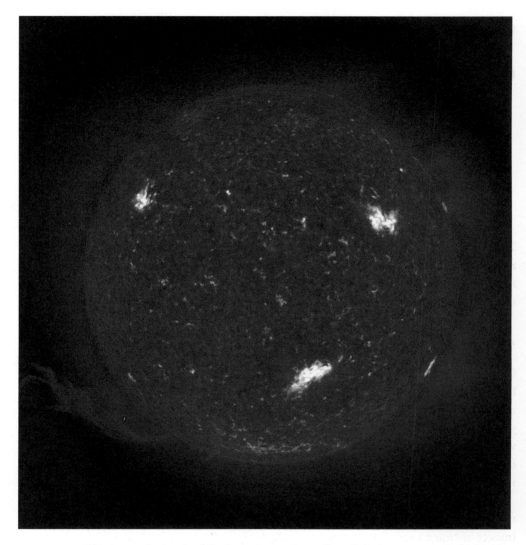

This impressive image of the Sun, a huge sphere of mostly hydrogen gas, was obtained in 1997 by the Extreme Ultraviolet Imaging Telescope (EIT) on SOHO. This spacecraft was launched in 1995 and placed at a permanent vantage point 1 million miles sunward of the Earth. The hottest areas appear white, while the darker red areas are at lower temperatures. Note the huge erupting prominence, many times the size of the Earth, at the lower left. These eruptions occur when a significant amount of ionized gas – atoms stripped of their electrons – escapes from the Sun's atmosphere and streams out into the interplanetary medium. When aimed in the direction of the Earth, powerful eruptions like the one pictured here sometimes produce major disruptions in the near-Earth environment, affecting communications, navigation systems and even power grids. The northern or southern lights – the Auroras – are caused by these storms. (SOHO/EIT Consortium. SOHO is a project of international cooperation between ESA and NASA.)

Chapter 1

Cooking the elements

> Far out in the uncharted backwaters of the unfashionable end of
> the western spiral arm of the galaxy lies a small unregarded
> yellow Sun.[1]

The origin of the energy from the Sun and stars was a long-standing mystery.
Hundreds of lifetimes were devoted to discovering the secret of the stars and,
in modern times, countless hours of computing time have been committed
to understanding the details of stars' lives. Now we know the answer.

Awakening

On a sunny summer day you probably enjoy spending an afternoon at the
beach, lying or sitting on pristine white sand, gazing at the unending blue
ocean in front of you. Although our Sun is far away, you can feel its heat on
your skin and will use a lotion to protect yourself from its damaging ultraviolet radiation, and wear dark glasses to relieve your eyes from its blinding
light. It is this energy that drives life on Earth. The light from the Sun is good
indeed, and you might wonder how it is produced.

If you go for a walk along the shore, you will find places where a great
variety of living things can be found: small fishes, crabs, algae, plants and
insects in "most beautiful and most wonderful forms." You will also find a
great assortment of the remnants of life, seashells, "some prettier than ordinary," which you may like to collect. Where did all this come from? At the
end of the beach there are rocks and some have very sharp hard edges. This
is evidently the product of erosion over a long time by the action of the waves
as they hit the shore. How long a time was needed? It was not until the twentieth century that scientists were able to provide answers to the above questions. You will learn about them as you read on.

By earthly standards the Sun is unimaginably far away, its typical distance from Earth being 150 million kilometers (93 million miles). Can you

---

[1] Douglas Adams, *The Hitchhiker's Guide to the Galaxy*, Six Stories, p. 5.

Above the photosphere of the Sun, the region from which sunlight is emitted, an atmosphere of very hot ionized gas extends for millions of miles: the chromosphere. This stunning ultraviolet image, obtained by the NASA Transition Region and Coronal Explorer (TRACE) spacecraft, shows fountains of million degree ionized gas flowing along solar magnetic fields forming giant arches, some over 300 000 miles high. (Standford–Lockheed Institute for Space Research)

imagine this? At 60 miles per hour it would take you about 175 years to reach the Sun, but light traveling at the enormous speed of 300 000 km (186 000 miles) every second takes only 8 minutes to traverse this large distance.

A fundamental law of physics states that nothing in nature can travel faster than light. It takes only about 0.1 second for light to circle the Earth, and light reaches the Moon in 1.25 seconds. The high speed of light, which is the speed of any electromagnetic wave, is the reason you can talk on the phone halfway around the Earth and get what seems to be an instant reply.

The diameter of the Earth is about 13 000 km (8000 miles), so the Sun is 11 500 Earth diameters from us. If the Earth were the size of a penny, the Sun would be a ball 7 feet across placed about 700 feet away. At this scale, the Moon, actually at a distance of 384 000 km (238 000 miles) from the Earth, would be almost 2 feet away and only ³⁄₁₆ inch in size.

The Earth completes an orbit around the Sun once per year, indeed this is the definition of a year. Today we accept this fact without batting an eyelid, but it was the cause of great turmoil in the sixteenth century, and caused much personal suffering to a few. An Earth that moved, and was not at the

"THERE WAS A TIME WHEN I THOUGHT THE EARTH REVOLVED AROUND HER."

© Nick Downes

center of the Universe created by God, was not in accord with Christian doctrine of that time, and to rock the boat was to risk severe punishment. For more than 1500 years the Christian doctrine had been in harmony with the views of nature proposed by Plato, Aristotle, and Ptolemy. In this view, all cosmic objects were attached to impenetrable transparent crystalline spheres and orbited the Earth in perfect circles at constant speed. They were composed of a special "perfect" fifth element – quintessence – that gave them perfection of form and let them obey laws of motion different from those of Earth. Nearest was the Moon (which does orbit the Earth, although not in a perfect circle and certainly not on a crystal sphere), followed by Mercury, Venus, the Sun, Mars, Jupiter and Saturn, with an eighth sphere occupied by the eternal, unchanging, stars. The outer planets, Uranus, Neptune, and Pluto had not yet been discovered.

This Earth-centered system, called geocentric, was outlined in one of the most influential books of antiquity, *The Almagest* ("the greatest") published in the ninth century AD by Arab translators of the original Greek work of Claudius Ptolemy. He was the most renowned astronomer and geographer of the ancient world, living at Alexandria in Egypt in the second century AD. To fit the observed positions of the planets, which sometimes appear to reverse their direction of motion in the sky, the Ptolemaic system stated that the planets were on circles, called epicycles, whose centers moved on other circles, called deferents, which were centered on the Earth. Since in this system Venus never reached a point where it was opposite the Sun from Earth, it would never show a full phase, like the full Moon, which we see when the Moon and the Sun are on opposite sides of the Earth. The Ptolemaic system was a complex, gigantic, clockwork mechanism pushed along by divine forces, and it described the observed motions of the planets and stars adequately.

Eventually, Nicolaus Copernicus rocked the boat with the publication of his book *De Revolutionibus Orbium Coelestium* (*On the Revolutions of the Heavenly Orbs*) in 1543, fifty-one years after Columbus reached America. It is said that a copy of his book was brought to Copernicus on the last day of his life, May 24, 1543, at Frauenburg, now Frombork, in Poland, on the coast of the Baltic Sea. In academic circles there is a saying "publish or perish," expressing the need for academics to publish research if they want to advance academically. For Copernicus we could say that it was a case of publishing *and* perishing. Copernicus was born at Torun, in Poland, on February 19, 1473, and became famous throughout Europe for his studies in astronomy. He proposed a system with the Sun at the center, a heliocentric system, which is our modern view. The heliocentric system was not really new. We know that Aristarchus, a Greek philosopher of the Pythagorean school, born about 310 BC, had formulated just such a system. However, we know of this only indirectly and the influence of Aristotle and Ptolemy put the heliocentric system on ice for 1500 years. In this system, the Earth rotated on its axis once per day, and moved about the Sun in a circular orbit once per year.

Although today we refer to the "Copernican Revolution," Copernicus was not a revolutionary. He delayed the publication of his book for many years because he feared rejection of his theory. The Copernican Revolution crept in through the back door, so to speak, over a period of many years. It was only half a century after the death of Copernicus that the heliocentric system became the subject of great controversy. For many, the problem with this new system was that the status of the Earth was downgraded to merely that of another planet, no longer the center of the Universe, contradicting the holy scriptures. The Catholic Church did not object to the heliocentric system as long as it was presented as a hypothesis to "save the appearances,"

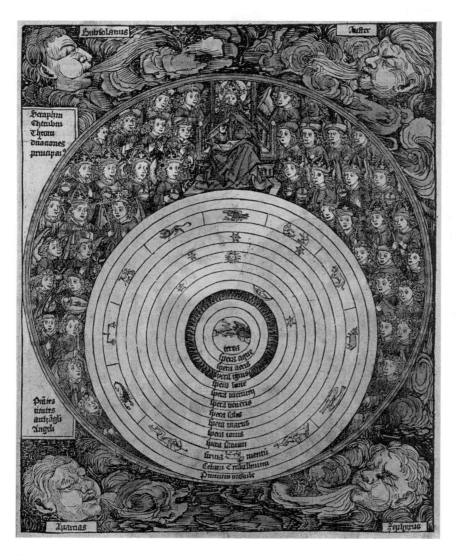

The Ptolemaic (geocentric) system is depicted here as part of an account of the last day of creation by Hartmann Schedel (1440–1514) in the Nuremberg Chronicles (Nuremberg, 1493). After the spheres of earth, air, water, and fire there were seven spheres with the wanderers – the planets – followed by the sphere of the fixed stars, and then the Primum Mobile responsible for imparting motion to the entire system. God on his throne views his creation accompanied by the nine choirs of angels (from the Seraphims to Angels) labeled in Latin on the left. The four winds are at the four corners. The reason that our week has seven days, and the names for the days of the week, are a remnant of this cosmic conception. (Courtesy of Adler Planetarium and Astronomy Museum, Chicago, Illinois)

"SURELY YOU WERE AWARE WHEN YOU ACCEPTED THE POSITION, PROFESSOR, THAT IT WAS 'PUBLISH OR PERISH.' "

© Nick Downes

that is, to agree with astronomical observations. In the heliocentric system, the order of the planets, also in circular orbits about the Sun, became as we know it today, Mercury being closest to the Sun followed by Venus, the Earth (with the Moon in orbit about it), Mars, Jupiter and Saturn. Still, the heliocentric system, based on circular orbits and motion at constant speed, was not accurate enough when used to predict the positions of the planets, which actually move in elliptical orbits.

Galileo Galilei, probably the best-known figure in the history of science,

The Copernican (heliocentric) system as depicted in *Harmonia macrocosmica* by Andreas Cellarius (Amsterdam, 1661).The Moon goes around the Earth, and Jupiter is shown with its four moons. (Courtesy of Adler Planetarium and Astronomy Museum, Chicago, Illinois)

was born in Pisa, Italy, on February 15, 1564, three days before the death of the great Michelangelo Buonarroti (1475–1564). He is considered the father of physics, having introduced the idea that the book of nature was written in the language of geometry and mathematics, together with the notion that to understand nature it is necessary to observe and experiment. Although his telescope was thousands of times less powerful than today's large optical telescopes, he made important discoveries with it, which he promptly published in 1610 in *Sidereus Nuncius* (*Sidereal Announcement*). In the book Galileo reported mountains on the Moon, which previously had been thought of as a perfect sphere made from quintessence. He also reported on innumerable "fixed" stars in the Milky Way, and most significantly announced the discovery of the moons of Jupiter, which he found to orbit this giant planet. This immediately showed that not everything revolved about the Earth as predicated by the geocentric theory of the Universe. Soon Galileo also observed the phases of Venus, including a full phase, which directly contradicted the geocentric model. He insisted that the Copernican system was not just a mere hypothesis to save the appearances, but the absolute truth, and the scriptures needed a new interpretation. This, and his acerbic attacks against those who differed with him, including members of the Jesuit order, led to his troubles with the ecclesiastic authorities.

Although Galileo did not realize it, he must have seen the then-undiscovered planet Neptune, since we can compute that in January of 1613 it lay close to Jupiter. As it was, Neptune was discovered much later, at the Berlin Observatory on September 23, 1846, by Johann Gottfried Galle (1812–1910) while searching a position in the sky predicted by the Englishman John Couch Adams (1819–1892) and the Frenchman Urbain Leverrier (1811–1877) to contain a new planet. These two mathematicians had independently computed this to explain observed irregularities in the orbit of Uranus, the seventh planet, discovered in England in 1787 by the German-born astronomer William Herschel (1738–1822). Adams and Leverrier correctly interpreted the irregularities of Uranus as a gravitational effect from a yet unknown eighth planet.

In Rome, in the year 1616, the chief theologian of the Roman Catholic Church, Roberto Cardinal Bellarmino (1542–1621), informed Galileo that the Sacred Congregation of the Holy Office had condemned Copernicanism as altogether opposed to holy scripture, and the book of Copernicus was placed in the *Index Librorum Prohibitorum*, the blacklist of books banned by the Inquisition. He notified Galileo that the Inquisition had prohibited defending the Copernican hypothesis or even believing it to be true.

In 1625, in Florence, Galileo began work on his *Dialogo Sopra I due Massimi Sistemi del Mondo, Tolemaico e Copernicano* (*Dialogue Concerning the Two Chief World Systems, Ptolemaic and Copernican*). It was published in 1632, causing a storm, although it had been cleared by ecclesiastic authorities. Maffeo Cardinal Barberini (1568–1644), elected Pope Urban VIII in 1623, was an old friend and supporter of Galileo, but felt betrayed by him, because his Dialogue clearly favored the Copernican system and appeared to make fun of Urban. At the end of 1632 Galileo was summoned to Rome to stand trial, at age 70 and in poor health, on *suspicion* of heresy. His book was found to favor the Copernican system and was added to the infamous index, where it remained until 1822. Galileo was pronounced guilty and made to humiliate himself by publicly abjuring his opinions, which he did on this knees before his judges at the Dominican convent adjacent to the church of Santa Maria Sopra Minerva, on June 22 of 1633. He was placed under house arrest in Arcetri near Florence for the rest of his life, which ended on January 8, 1642. His remains are in a tomb, built only in 1737, at the church of Santa Croce,

In this view, the Sun is setting over the cloud-covered Pacific Ocean behind the European Southern Observatory's Very Large Telescope Array (VLT) at the summit of Cerro Paranal, at an altitude of 8600 feet (2635 meters) in the northern Chilean Andes mountains. The Paranal area in the Chilean Atacama desert is believed to be the best site for astronomical observations in the southern hemisphere. Four telescopes with gigantic 26 feet (8 meter) diameter mirrors can work in single or in combined mode. In the latter, the VLT provides the total light-collecting power of a 16-meter single telescope. The YEPUN (*Sirius* in the Mapuche language) telescope is in the front, with the main mirror cover in place. From left to right in the background are, Antu (*The Sun*), Kueyen (*The Moon*) and Melipal (*The Southern Cross*). In his *Sidereus Nuncius*, Galileo wrote: "other things, possibly more important, will with time be discovered by me or by others, with the aid of similar instruments . . ." The VLT is a marvel of modern science and engineering – Galileo would be proud of us. (ESO)

in Florence, near that of Michelangelo. As we shall see later, even worse things could happen in those days to those who did not toe the line.

There is no evidence that Galileo said under his breath "eppur si muove" (yet it does move) after his trial, when he formally renounced the view that the Earth moves, although I can imagine that it crossed his mind many times. During the last years of his life he wrote his most important treatise,

Discorsi e Demostrazione Matematiche Intorno a Due Nuove Scienze (*Discourses and Mathematical Demonstrations Concerning Two New Sciences*), smuggled out of Italy and published in Leyden in 1638, which laid down the foundation of the science of mechanics. It is ironic to think that, had he not been embroiled in this miserable episode and prevented from further studies in astronomy, he would perhaps have never returned to physics. Three hundred and sixty years after Galileo's conviction, the Catholic Church reviewed his trial and found that, because he had not been convicted of heresy by the Inquisition, there was no need to acquit him.

On December 25 of the year of Galileo's death, as if passing the baton, Isaac Newton (1642–1727), possibly the greatest scientist of all times, was born in Woolsthorpe, Lincolnshire, England. The publication in 1687 of his *Philosophiae Naturalis Principia Mathematica* (*Mathematical Principles of Natural Philosophy: Principia* for short) represents the culmination of the Copernican revolution, and stands at the foundation of physical science. It is, without doubt, the most important work ever published in the physical sciences. As he himself acknowledged, Newton published this work after being encouraged and helped by Edmund Halley (1656–1742), after whom the famous comet is named. In his book, Newton formulated the Law of Universal Gravitation expressing the way any object exerts an *attractive* force on every other object, and the laws of motion.

The gravitational force between the Earth and the Sun determines the Earth's orbit. The same is true of all the planets. The orbit of the Moon around the Earth is similarly determined by the force of attraction between them. The gravitational force is large for large masses, and becomes larger when two objects are close to each other. It varies as the inverse of the square of the distance between the two masses, so that if we move 10 times closer to a mass, the force becomes 100 times stronger. Gravitation is what keeps our atmosphere from drifting away from the Earth into space. It is because of gravity that when we lie on the beach we do not fall off, but cling to the surface of our planet like magnets to a refrigerator door. You might then ask how it is that the Moon, attracted by the Earth, does not fall on us? It is precisely because it is in orbit around the Earth that it does not, the gravitational force being balanced by the centrifugal force (the same applies to spacecraft in orbit). This is also what keeps the Earth from falling into the Sun. Should the Earth suddenly stop in its orbit – don't worry it cannot happen – then it would fall toward the Sun with ever-increasing speed, to be consumed after a voyage of only three months' duration. The Earth attracts each of us toward its center and this is what we call our weight. We do not fall toward the center because the ground prevents this, except when we have no ground to stand on, usually an uncomfortable position to be in. It became possible, using Newton's laws of motion and gravitation, to compute and predict the observed motions of all objects on Earth and in the solar system. In this way, Newton unified the heavens and the Earth, as there was no longer a contrast between a perfect heaven ruled by eternal and immutable laws and the imperfect and changing Earth, nor a need for divine movers.

Newton wrote, "If I have seen farther than other men, it is because I have stood on the shoulders of giants," acknowledging the work of his predecessors. He died in 1727 and was laid to rest in Westminster Abbey. Shortly before his death he declared, "I do not know what I may appear to the world, but to myself I seem to have been only like a boy playing on the seashore, and diverting myself in now and then finding a smoother pebble or a prettier shell than ordinary, whilst the great ocean of truth lay all undiscovered before me." The physical laws presented in Newton's *Principia* form the basis of mechanics and gravitation, and have provided the tools which over the past 300 years have led to a greater understanding of "that great ocean of truth." Engineering calculations to build bridges and airplanes, and the computations to send the Apollo astronauts to the Moon, are all based on Newtonian physics.

A star called Sun

Throughout this book we shall need to talk about huge quantities, great distances, long times, and large populations. We measure these in millions (1 000 000) and billions (1000 million, or 1 000 000 000). Distances in the cosmos are so large that it does not make sense to specify them using the common yardsticks we use to measure distances on Earth, such as miles or kilometers. Even billions of these will not get you very far in the Universe. We can specify large distances by the time it takes light to travel them at its enormous speed, allowing us to use smaller numbers. Thus we can say that the Sun is 8 light-minutes, instead of 93 000 000 miles, away. The distance from the Earth to the Sun is also called 1 astronomical unit (AU). It takes light about 5 hours to reach us from distant Pluto, the outermost planet of the solar system, so we say that it is at a distance of about 5 light-hours, or 40 AU, from the Sun.

Looking at the night sky you can see hundreds of points of light. Most of these are stars in our galaxy: the "Milky Way," one of perhaps billions of galaxies in the Universe. The nearest star to the Sun is Proxima Centauri, the smallest member of the triple star system of Alpha Centauri at a distance of

We live inside the Milky Way galaxy, so called because to us it appears as a whitish glowing band in the night sky. It is a huge disk-shaped system composed of billions of stars. Our Sun is just one of them. When this galaxy is photographed with a long exposure, features not perceived with the naked eye become visible. Here, a part of it, towards the galactic center in the Sagittarius constellation, is shown. Dark bands of interstellar dust are visible against the light of millions of stars, some of them in clusters. Red emission nebulae and blue reflection nebulae are also visible. The Milky Way would look like the Andromeda galaxy (page 15) if viewed from far away. Note that this image looks like a blow-up of a small piece of the spiral galaxy NGC 891, shown on page 16. Antares, a red supergiant star 300 times larger that the Sun, is the bright star visible at bottom center. (JPL/NASA/Table Mountain Observatory. Photograph by James W. Young.)

4.2 light-years. One light-year, the distance traveled by light in one year, equals 5878660000000 miles, so you can see why using miles to talk about cosmic distances is not practical. On the scale of the model where the Sun was 700 feet from Earth and the Earth was the size of a penny, Proxima Centauri would be at a distance of 35000 miles. As you can see, "near" is in this case "nowhere near." In case you did not realize it, the Sun *is* a star, quite an ordinary one at that, and obviously the nearest one to Earth. The Milky Way is a gigantic disk-shaped system composed of stars, interstellar gas, dust, and mysterious dark matter. Astronomers estimate that it contains some 200 billion stars, more than the number of grains of sand on a beach several miles long. Light takes 100 000 years to travel from one end of its disk to the other, so if you find that the Sun is unimaginably far away do not even try to imagine galactic distances. Other galaxies are millions and even billions of light-years away. What can I say? *The Universe is an astonishingly, almost impossibly large place.* (And it's getting larger!)

Our Sun is found in what we could call the "suburbs" of the Milky Way, about 30 000 light-years from its center. It orbits the center of the Galaxy, under the influence of its gravitational pull, once in every 250 million years, a quantity we could call a galactic year. If you observe the sky carefully from a dark location, far away from city lights, you will notice that most stars concentrate in a broad band of diffuse light across the sky. This band is our Milky Way galaxy as seen from our location within it. When observed with binoculars or with a telescope, even a small one such as that used for the first time by Galileo in 1609, the diffuse light is seen to be coming from millions of stars too faint to be distinguished individually with the naked eye.

The gaseous surface of the Sun, called its photosphere, is at a temperature of about 6000 degrees Celsius (11 000 degrees Fahrenheit). This is quite high, about six times hotter than the temperature of lava in a volcanic eruption. Even though we are at a large distance from the Sun, we feel this heat, because of the Sun's large size, its diameter being about 110 times that of Earth. In other words, the furnace is not only quite hot but it is gigantic. From the Sun's perspective, the Earth is only 115 Sun diameters away, quite *near* compared with typical distances between stars, which are tens of millions of Sun diameters apart.

The Sun's energy output, its luminosity, is equivalent to an enormous 400 trillion trillion watts (a trillion trillion is a 1 followed by 24 zeros). As in many astronomical situations, this quantity is not something we can immediately comprehend. After all, if it were only 23 zeros, it would not be much different in terms of our understanding of the quantity. However, life on Earth would be quite upset about the missing zero since it would not survive this lowering of the Sun's energy output by a factor of ten. Maybe you can get a better notion of the size of this quantity if I tell you that the Sun generates more energy in one second than has been consumed in the whole history of humanity. The Sun radiates its energy into space uniformly in all directions and our planet intercepts only a tiny fraction, about one part in

Clusters of galaxies are the largest structures in the Universe. They consist of hundreds or thousands of galaxies moving in orbits about the central part guided by the gravitational force due to their collective masses. This image, obtained with the 3.5-meter WIYN telescope on Kitt Peak in Arizona, shows the Abell 98 cluster of galaxies. It is located at the enormous distance of over 1000 million light-years from us, meaning that the light that made this image traveled for 1000 million years to get from there to here. Almost every speck of light in this photograph is a galaxy, some being spherical and seen as round images, and others being disks which sometimes appear flat. Hundreds of galaxies are part of the cluster, each containing billions of stars. (Mike Pierce (Indiana)/WIYN/NOAO/NSF)

The majestic Andromeda spiral galaxy with its bright nucleus can be seen as we peer into our Universe through hundreds of foreground stars of the Milky Way. It is the nearest large galaxy at a distance of "only" 3 million light-years. The light which made this image left Andromeda 3 million years ago, when our species was nowhere to be found. Many things happened on Earth as the light traveled toward us, finally to enter the tube of the telescope that made this image, situated on Kitt Peak. Billions of stars produce the diffuse light against which dark bands of interstellar gas and dust can be seen. This disk measures about 100 000 light-years in diameter. Also visible in the image are two small companion galaxies: M32 at the lower center and NGC 205 at the upper right. These two galaxies slowly orbit Andromeda like moons about a planet. The Milky Way also has two small satellite galaxies: the Large and the Small Magellanic clouds. (Bill Schoening, Vanessa Harvey/REU program/AURA/NOAO/NSF)

This is the spiral galaxy NGC 891 (number 891 in the "New General Catalogue of Nebulae and Star Clusters" which appeared as Volume 49, Number 1, of the *Memoirs of the Royal Astronomical Society* in 1888). At a distance of 10 million light-years, it is oriented so that we see it edge-on, and see the disc of interstellar dust against the background light of its many stars. The bright stars visible on the photograph are in the foreground and belong to the Milky Way. Astronomers have catalogued millions of galaxies, many farther than a billion light-years. (Blair Savage, Chris Howk (U. Wisconsin)/N.A.Sharp (NOAO)/WIYN/NSF)

2000 million (2 billion). It is this energy that supports life on Earth. Although the Sun emits some of its energy as invisible radiation – radio waves, infrared, ultraviolet, and X-ray radiation – it produces most of it as visible light, which is of course what we can see. Ultraviolet light is dangerous to life but most of this is fortunately intercepted by a thin protective layer of ozone in the stratosphere.

All stars are basically the same thing: large luminous spheres of gas. However, there is a large variety of stars from small ones with only one-tenth the mass of the Sun and one-thousandth of its luminosity, such as the nearby Proxima Centauri, to giants with 50 times the solar mass and 50 000 times its luminosity, such as Betelgeuse in the constellation Orion. Our Sun is larger than most. Stars appear faint because they are so much farther from us than the Sun, just as a bright light appears faint when viewed from a large distance. Because of the Sun's proximity, we can study it in great detail, providing us with the key to understanding the stars.

We have observed and speculated about the nature of stars throughout history, but only in the twentieth century did it become possible to understand their structure, evolution and, in particular, the source of the enormous energy they produce. A star contains mostly hydrogen (about 75 percent of the mass) and helium (about 25 percent), and small amounts of the heavier elements. In fact, helium was first discovered by astronomers studying the Sun, who determined that a then-unknown element was present, naming it helium (from the Greek *Helios*, "Sun"). The gas of a star forms a sphere in equilibrium between the gravitational force of its own mass trying to collapse it toward its center, and the pressure of the hot gas pushing outward. One hundred years ago it was impossible to explain how the Sun's great luminosity could be sustained over a time span of at least 4.5 thousand million (4.5 billion) years, the age of the solar system obtained early in the last century from the radioactive dating of rocks and meteorites. The theory of relativity, formulated by Albert Einstein (1879–1955), and his famous relation, $E = mc^2$, provided part of the answer.

Atoms and energy

Einstein, who stands with Newton at the pinnacle of human thought about our physical Universe, was born in Ulm, Germany, on March 14, 1879. Upon graduation from the Swiss Federal Institute of Technology in 1900, he finally landed a job as a technical expert at the Swiss patent office in Bern. He remained at this job until 1909, and during this time obtained several important results which would revolutionize theoretical physics. Newtonian physics became a subset of Einstein's much wider and more fundamental conception of our natural world, as laid out in his theory of relativity and his theory of gravitation. Einstein was awarded the 1921 Nobel Prize in Physics, interestingly, not for his new theories, but for the explanation he worked out for the photoelectric effect, which was based on the also new and revolutionary ideas of quantum mechanics. In 1921, the world of physics had not yet been able to understand the full and profound significance of Einstein's new theories.

The physics of atoms and nuclear reactions, developed early last century, provided the other part needed to understand the workings of the Sun and the other stars. Einstein's formula tells us that mass and energy are related in such a way that *if it were possible* to convert some mass (denoted by m in his relation) into energy (E in the relation) the quantity of energy obtained is computed by multiplying the mass by the square of the speed of light (c in the relation). Because, as we have seen, the speed of light is a very large number, its square will be even larger, so that even a minute amount of mass can be converted into an extremely large quantity of energy. The difficulty, of course, is that this will only work if a process can be found that will start with some mass and end with less mass, the difference being converted into energy. Since stars are mostly composed of hydrogen, it is reasonable to investigate whether a reaction involving this element can provide such a process.

But, before we look at this I need to explain a couple of basic things about atoms, the things out of which everything is made, including you and me. Atoms, the basic building blocks of all matter, are composed of a nucleus, made up of protons with a positive electrical charge and uncharged neutrons in comparable numbers, and a surrounding cloud of electrons. There are as many electrons as there are protons in the nucleus. Each electron has an equal and opposite charge to that of the proton, so that an atom, as seen from the outside, is electrically neutral. Almost all the mass of an atom is in the nucleus since electrons are about 2000 times lighter than protons and neutrons, which have almost equal masses. The number of protons (the so-called atomic number) in an atom's nucleus determines its chemical nature. As an example, an element with 26 protons is iron, the most common element on Earth, which normally also has 30 neutrons. Iron's atomic mass number, the total number of neutrons and protons in the nucleus, is 56 (26+30) and we denote this by a superscript in front of the symbol for the chemical element: ^{56}Fe. A rare atom with 77 protons is iridium (^{193}Ir), and one with 79 protons is gold (^{197}Au), quite common in jewelry.

Carbon, the most important element of life, has six protons and usually six neutrons (^{12}C) but there is a less abundant variety with seven neutrons (^{13}C) and one with eight neutrons (^{14}C). These are called isotopes of carbon. The isotopes of an element have identical chemical properties, since they have the same number of protons, but slightly different masses. Some isotopes are unstable and will, after some time, decay spontaneously into a different form. These isotopes are called radioactive, and they decay in such a way that half of the material will be changed after a certain time, called the half-life, has passed. As an example ^{14}C (carbon-14) will decay to ^{14}N (nitrogen-

14) with a half-life of 5730 years. In the process, a neutron in the nucleus is transformed into a proton, and an electron is emitted. As we shall see, knowing the decay properties for different radioactive atoms gives us a tool that allows us to learn the ages of fossils and rocks. We shall also see later that iridium, almost absent on Earth's surface but found in meteorites, is related to this story in a surprising way.

The periodic table of the elements is an arrangement of all the elements in order of increasing atomic number. There are 92 naturally occurring elements, from hydrogen to uranium (^{238}U), each element having one more proton in its nucleus than the previous one. Elements combine to form molecules, some as simple as water (H_2O), composed of two hydrogen atoms and one oxygen atom, or molecular oxygen, which we breathe, combining two oxygen atoms (O_2) (atomic oxygen is poisonous). Other molecules are so complex that they contain millions of atoms, like the molecule of ribonucleic acid (RNA), a very important component of the cell, as we shall see. The RNA molecule is a long chain composed of carbon, hydrogen, oxygen,

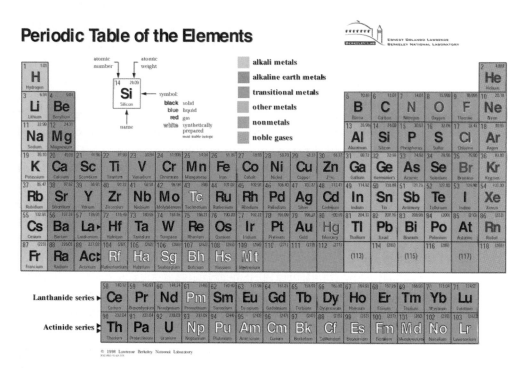

Periodic Table of the Elements

When the Universe came into existence some 15 billion years ago, the only elements produced were hydrogen, helium, and traces of lithium, beryllium, and boron. The heavier elements did not yet exist. They are the product of nuclear reactions inside stars and in supernova explosions.

nitrogen, and phosphorous atoms. Living things are mostly made up of just these few "biogenic elements," also including sulfur and calcium, with only traces of other elements.

So let us reflect for a moment. All the objects in our world, including you and I, all we eat and drink, and all the bewildering variety of things we observe, are the complex manifestation of an amazingly simple system of 92 different combinations of protons and neutrons. These elements, most of the time just a few of them, can form organic molecules containing carbon, the building blocks of all living things, or minerals, the building blocks of rocks. The bonds that hold atoms together in a mineral or a molecule are the result of electrical forces which arise from the different ways in which the electrons in the atoms can arrange themselves. The marvelous variety of phenomena we observe is mostly the result of different chemical reactions between these compounds, which combine, or break up, to form new compounds. These *chemical* reactions will sometimes require energy to happen, and sometimes release energy as they occur. The atoms involved in chemical reactions do not change; they just combine to produce different compounds. *Nuclear* reactions, on the other hand, change the composition of the atomic nuclei, and so do transform one element into another, as we shall see.

The Sun's secret

The nucleus of hydrogen (^{1}H), the simplest atom in nature, is composed of just a single proton, while the nucleus of helium (^{4}He), the next element in the periodic table, is composed of two protons and two neutrons. You might also remember from that physics course you had to take that "like electrical charges repel each other," and this will happen between protons. The force of repulsion becomes very strong as they approach each other. For nuclear reactions to happen, the protons and neutrons involved must come extremely close together, because it is only at exceedingly small separations that another force of nature, the *nuclear force of attraction*, operates. It is this nuclear force, the strongest force in nature, that holds the protons in an atomic nucleus together despite the electrical repulsion between them (since they are of like charges).

A look at the periodic table shows that the atomic weight (the mass specified in relation to one-twelfth the mass of ^{12}C) of hydrogen is 1.008, while that of helium is 4.003. You will notice that the mass of helium is smaller than the combined mass of four hydrogen nuclei (4.032) by $^7/_{10}$ of a percent (0.7 percent), so if it were possible to convert four hydrogen atoms

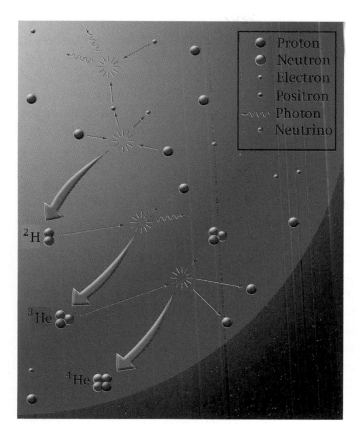

The fundamental process in a star. In the central hot core of a star like the Sun the nuclear fusion reactions converting four hydrogen nuclei into one helium nucleus proceed as illustrated. The first slow reaction converts two protons into a deuterium nucleus (^{2}H) releasing a neutrino and a positron in the process. Deuterium will rapidly combine with another proton to produce a light isotope of helium (^{3}He), which itself combines with a neutron from another helium-3 to finally produce normal helium (^{4}He). The positrons produced rapidly collide with electrons and annihilate, producing additional energy, while the neutrinos escape the Sun. Fortunately the helium-2 (two protons) nucleus does not exist in nature, otherwise this process would occur very rapidly and stars would not exist for long. (José F. Salgado)

into one helium atom, according to Einstein's formula, energy would be obtained. Well, you can guess the answer. It is indeed possible, and is the basis of the chain of thermonuclear fusion reactions occurring in the core of a star. It is only at very high temperatures that the charged nuclei involved travel fast enough to overcome the electrical force of repulsion, and get close enough to each other for the nuclear force to take over. So the fundamental requirement for this reaction to occur is extremely high temperature (the "thermo" in thermonuclear), and this happens at the Sun's center, where the temperature reaches an amazing 15 million degrees Celsius. This is a consequence of the enormous pressure generated by the large mass of the Sun, which is about 330 000 times the mass of the Earth. At this extremely high temperature, matter behaves like a gas although the density at the center of the Sun is about 20 times greater than the density of iron. If the mass of the Sun were about 10 times less, then the temperature at its center would not be sufficient for these reactions to proceed. So we do not expect stars with masses much less than one-tenth of that of the Sun to generate energy by hydrogen fusion, and in fact we have not found any that do.

The chain of thermonuclear fusion reactions, the details of which are shown in the figure, starts with the conversion of a proton to a neutron, producing along the way two other particles, a positron and a neutrino. The positron is a particle made of what is called antimatter. It is identical to the electron but has the opposite charge. When matter meets antimatter it is instantly converted into energy – it annihilates. Every particle in nature has its antiparticle. There are antiprotons and antineutrons, and with these you could build antiatoms and antimolecules. We can imagine an antiworld somewhere far away in another galaxy, although research has found no evidence for this. Viewed from a distance it would *look* the same as our world because the light by which we see things is the same for matter and antimatter. Anticars would look just like cars, and anticities just like cities. We would not be able to see that an antiperson was, well, an anitiperson. So Captain Kirk would be well advised, as he boldly goes where no man has gone before, to be extremely careful before shaking hands with strangers. At the Sun's center, the newly created positron will meet one of the many electrons present and instantly annihilate, contributing to the total energy output of the Sun.

The neutrino is an extremely elusive particle which barely interacts with matter. Therefore, most of those produced escape from the core of the Sun with only a very small likelihood of striking an atom in the Sun's body. Most of those intercepting Earth go right through it as well. Indeed, roughly 50 000 billion neutrinos from the Sun bombard your body every second, by day from above and at night from below, and if you were not reading this you would never know! Although unlikely to react, the huge number of neutrinos produced means that a few *will* react with matter and can therefore be detected. This is precisely what several experiments have set out to do, and by capturing a few neutrinos they have confirmed directly that such reactions are taking place at the center of the Sun.

This first reaction in the chain takes very much longer than the following ones, by a factor of 10^{18} (a 1 followed by 18 zeros), and so the entire reaction chain is slow. If it were not for this, the Sun and all the stars would have burnt out explosively long ago. After the slow reaction that produced them, neutrons can rapidly combine with a proton (no repulsive force here) to form the nucleus of deuterium (^{2}H), a heavy isotope of hydrogen. Deuterium will rapidly combine with another proton to produce a light isotope of helium (^{3}He), which itself combines with a neutron from another ^{3}He to finally produce normal helium (^{4}He). The dreadful hydrogen bomb is a device which uses the same reactions as those in the Sun, but starts with the heavier isotopes of hydrogen – deuterium and tritium (a proton with two neutrons) – and so these proceed explosively. To generate the necessary high temperatures, a uranium-based atomic bomb is used as a "trigger."

The 1967 Nobel Prize in Physics was awarded to German-born Hans Albrecht Bethe (1906–) for his contributions to the theory of energy generation in stars.

An intriguing detail about the strong nuclear force is that it is not strong enough to allow just two protons to stick together, and an additional neutron is needed to contribute to the nuclear force for this to happen. This means that the light isotope of helium (^{3}He) is stable but that an even lighter isotope ^{2}He (two protons) does not exist in nature. This is rather fortunate since, if this were not so, fusion reactions in a star would not be slowed by the need to first produce a neutron. What is remarkable is that, if the nuclear force were only 4 percent stronger, ^{2}He would be stable and you would not be reading this! There are many other examples in nature where we find that if a fundamental quantity were to have a slightly different value then the Universe would be very different, so different that the story I am telling you would not have happened. This might be of profound significance, or may simply mean that, if things were not the way they are, we would not be here to find it profound.

Every second, the Sun converts 5 million tons of matter into energy at its center, and so every second the Sun becomes that much lighter. During its 4.5-billion-year history the Sun has converted a gigantic amount of mass, equivalent to about 100 Earth masses, into energy. This is, however, only $\frac{1}{3000}$ of the Sun's mass and does not significantly affect it.

We know that, on Earth, intelligent life has taken 4 billion years to develop to the point where I am here to write about it. Fortunately, a star like the Sun is in *stable* equilibrium, a simple mechanism avoiding a premature end to its (and the Earth's) existence. This relates to the general property that if you compress a gas, its temperature increases, while if you expand a gas it cools. The rate at which fusion reactions occur in the hot core of the Sun, or any other star, depends on the temperature in such a way that an increase in temperature will increase the rate drastically and vice versa. Should the temperature at the core rise for any reason, the increased energy output from the core would increase the pressure and the core would expand. Consequently the temperature would be lowered and the reaction rate would slow down. On the other hand should the temperature decrease, the reaction rate would decrease and so would the pressure. The core would then begin to contract, increasing the temperature and restoring equilibrium. This self-regulation acts as a thermostat and will work as long as the reactions can be maintained, which for the Sun is for about 10 billion years, as explained in the boxed text. The result of this simple calculation – don't be alarmed; it is the only one that I will show you in this book – is no different from that of far more sophisticated calculations in which many details are included. Thus we see that our Sun is about halfway through its life and will be around for another 5 billion years or so, something that I am sure will allow you to sleep better tonight. Although the numbers are different for different stars, the fundamental conclusion holds: *stars are not forever* (neither are diamonds). *They are born, evolve and die.*

The life of the Sun

The basic nuclear reactions at the center of the Sun convert four nuclei of hydrogen (protons) into one helium nucleus composed of two protons and two neutrons, with the release of neutrinos and energy as summarized in the following relation:

Knowing some values from laboratory measurements, it is surprisingly easy to arrive at the lifetime T of the Sun. Here is how it is done:

About 10% of the hydrogen in the Sun is at sufficiently high temperature near its center (15 million degrees) for this reaction to proceed and convert hydrogen to helium. The lifetime T is obtained by taking the total amount of available "fuel," in this case 10 percent of the mass of the Sun (2×10^{29} kilograms), multiplying this by the energy obtained from the reaction for each kilogram, which we know from laboratory measurements (6×10^{14} joules per kilogram), and dividing this by the rate at which this energy is radiated into space by the Sun (its luminosity: 4×10^{26} watts, or joules per second). So we get

$$T = \frac{2 \times 10^{29} \text{ kilograms} \times 6 \times 10^{14} \text{ joules per kilogram}}{4 \times 10^{26} \text{ joules per second}} \sim 3 \times 10^{17} \text{s} \sim 10 \text{ billion years}$$

So the lifetime of the Sun is about 10 billion years.

Although stars of greater mass than the Sun have more hydrogen fuel available, their central temperatures are so much higher that they burn it much faster and so live for a shorter time. A star of about 10 solar masses lasts only tens of millions of years, a mere instant of cosmic time. Those we observe today must therefore have been formed "recently," leading astronomers to search for their birthplaces and their remains. In contrast, low-mass stars, of less than a half as much matter as the Sun, will live for tens of billions of years.

The future

So what happens to a star once it has used up all the hydrogen in its core? The answer to this question depends upon the mass of the star. What is left is a core full of helium atoms and no energy generation to maintain internal pressure. The equilibrium which kept the star stable for so long will be lost. Under the inexorable force of gravity, the star will rapidly contract until the central temperature rises to a level at which the fusion of helium can start, producing even heavier elements. This higher temperature, about 100 million degrees Celsius, is necessary to overcome the repulsive forces between helium nuclei, which have double the electric charge of hydrogen. There will be a shell, surrounding the core, where hydrogen fusion will be established because of the increased temperature there. The overall increased energy output at the star's center will lead to expansion of its outer layers, and it will increase in size by a large factor, becoming what is known as a red giant, red because the expanding outer layers will cool and appear redder, giant because this is what it is.

Three helium atoms can fuse to produce carbon, and this can then fuse with a further helium atom to produce oxygen, neon and magnesium. If the star is massive enough, its center can reach even higher temperatures, and it can continue past the fusion of helium to produce even heavier elements by the fusion of carbon and oxygen, leading to silicon, nickel and iron. All these fusion reactions will not add much to the star's lifetime because they happen at a rapid rate, much faster than the initial hydrogen fusion stage (which astronomers call the Main Sequence lifetime of the star). At these extremely high temperatures, many different reactions are happening simultaneously, with nuclei hitting each other, and colliding with protons and neutrons traveling at high speeds, sometimes breaking up in one collision only to be fused with another nucleus the next instant. *A star is a gigantic cauldron where some of the elements of the periodic table are "cooked."*

At some point, a sufficiently massive star will be composed of several

The Ring nebula in the constellation Lyra is a cylindrical shell of glowing gas seen almost end-on, which is why it looks like a circle. It is about 2000 light-years from us and has a diameter of about one light-year. The central white dwarf star is what remains of the doomed star, which had more mass than the Sun. If, 6 billion years from now, there is "someone" looking in our direction, "it" might see something like this, displaying the death of the Sun. (Hubble Heritage Team AURA/STScI/NASA)

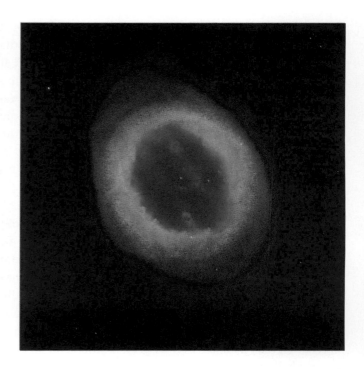

concentric shells, each dominated by a particular fusion reaction, the hottest shell at the center containing iron. The star will have increased its size by factors of several hundred, its energy output by factors of many thousand, and its surface cooled down to where it looks red. Betelgeuse, the second brightest star in Orion, is such a supergiant star with a surface temperature only about one half of that of the Sun, but about 50 000 times more luminous and so large that, if placed where the Sun is, its surface would reach the orbit of Jupiter. If you look at it, you can see that its color is red.

Our Sun, being of low mass, cannot go past the helium fusion stage. Detailed computer modeling of the Sun's evolution show that its luminosity was 30 percent lower when the solar system formed and has since increased steadily. This slow increase will continue into the future and in 1 billion years time the 10 percent brighter Sun will cause water vapor on Earth to be lost into space. Any large life forms on Earth's surface will face an unprecedented environmental crisis and succumb. The luminosity of the Sun will continue to increase and at some point will cause the evaporation of the oceans and scorching of the Earth. So much for global warming! Our planet will be sterilized.

Still, our species has been on this planet for only a couple of hundred thousand years, so 1000 million years is a very long way to go. There will be many other hurdles to overcome before worrying about this ultimate one.

Once helium fusion starts in another 5 billion years, it will take a relatively short half-billion additional years for the Sun's luminosity to increase by a factor of 1000. The surface of the red giant Sun will then be close to the orbit of our planet, and it will loom like a gigantic red disk filling half of the sky. At the same time, the Sun will be spewing a large amount of gas into space, losing a significant fraction of its mass, so that its gravitational pull on Earth will become smaller. Thus, the orbit of the Earth will become larger. Should the expanding, and sometimes oscillating, Sun reach the orbit of the Earth, then, because of the friction with the Sun's atmosphere, it will slowly spiral into it, and into oblivion. The world will end in fire. But should the Earth escape from the Sun as it moves outward, then it might survive this fiery death, and as the cooling small remnant of the Sun retreats at the center, it will remain forever frozen with no memory of things past. The world will end in ice.

The Magellanic Clouds are named after Fernao de Magalhaes (c. 1470–1521) the Portuguese captain general who set out to circumnavigate the Earth. Magellan sailed from the port of Sanlúcar de Barrameda, the outport for Seville, Spain, leading a fleet of five ships with a total crew of about 250 on 20 September, 1519. Surviving several attempts at mutiny, and the ravages of hunger, scurvy, and cold, he found a passage from the Atlantic to the Pacific Ocean, today known as the straight of Magellan, at the southern end of South America. His adventure continued until he arrived at the Philippines where he foolishly became embroiled in a local dispute and on April 29, 1521, got himself killed on the island of Mactan. History records that Juan Sebastian de Elcano, commanding *Victoria*, the only surviving vessel, arrived back to Seville with a crew of just 18 on September 6, 1522, three years after having departed. The rest of the party had either died of starvation and disease, or been killed in different skirmishes. The survivors were able to attest beyond any doubt: *our Earth was round indeed.* At this pace it would have taken Magellan 30 years to reach the Moon, whereas the Apollo astronauts reached it in 66 hours. In this painting, *Victoria* is seen traveling toward the Magellanic clouds, a journey that would have taken her longer than the age of the Universe, estimated to be 15 billion years. (Courtesy of the Morrison Planetarium. Art by Lynette R. Cook)

Chapter 2

The fertilization of space

The best bang since the big one[1]

The fate of a star, once its leisurely life of hydrogen fusion ends, depends on its mass. Some stars end their lives violently and others fade away quietly.

Gold and platinum

In the eighteenth century, the French astronomer Charles Messier (1730–1817) compiled a catalog of nebulae, objects which appear fuzzy when viewed through a telescope. Comet hunting was, and still is, a way to instant fame, and Messier compiled his catalog so that astronomers would not waste their time by confusing these nebulae with comets. The first entry in his catalog, M1, was in the constellation Taurus. Little did Messier suspect, as he observed M1 in 1758, that it would become one of the most extensively studied objects by modern astronomers. The Irish astronomer William Parsons (1800–1867), the third Earl of Rosse, and builder of the largest telescope of the nineteenth century, observed the M1 nebula around 1850. The nebula got its name, the Crab nebula, because (with some imagination) it resembles a crab on Parsons' drawings.

From ancient times, observers of the sky have noticed that stars do at times flare-up, becoming very bright for a short time, contradicting the view that stars are immutable. Sometimes a star might appear for a few weeks or months where none had been visible before. Such stars were called novae. This was also observed to happen to stars known to be at enormous distances from us, perhaps even in another galaxy, indicating that in these cases the increase in luminosity clearly had to be enormous, much more than for a nova, leading to the name supernova.

In the year 1054, Chinese astronomers recorded their astonishment at

[1] Douglas Adams, *The Hitchhiker's Guide to the Galaxy.*

(a)

One thousand years after the explosion of a massive star, its debris has expanded to fill a vast region with a diameter of about 15 light-years. Above: this spectacular image of the Crab nebula in the Taurus constellation was obtained by the European Southern Observatory's Very Large Telescope (VLT). For 5000 years the flash of the explosion traveled to us, and upon arrival at Earth, was observed in the year 1054 by Chinese astronomers. The different colors recorded in the light from the Crab nebula are produced by different processes taking place in the gas and dust expelled by the explosion. The core of the star has survived the explosion as a "pulsar," which is the lower of the two moderately bright stars visible to the upper left of center in the Hubble Space Telescope image (b). The energy from this neutron star, which spins on its axis 30 times a second, heats the surrounding gas creating the ghostly diffuse bluish-green glowing gas cloud in its vicinity, including a blue arc just to the right of the neutron star. The colorful network of filaments is the material from the outer layers of the star that was expelled during the explosion and is now expanding outward at high speed. The image (c) shows a sequence, only 2.5 seconds long, of very short exposures of the two central stars. You can see one of them pulsing, whereas the light from the other star (and a second fainter one) does not change. This is one of the few pulsars that can be seen to pulse optically, since most pulse in radio waves. (Main image ESO; (b) STScI/NASA; (c) ESO)

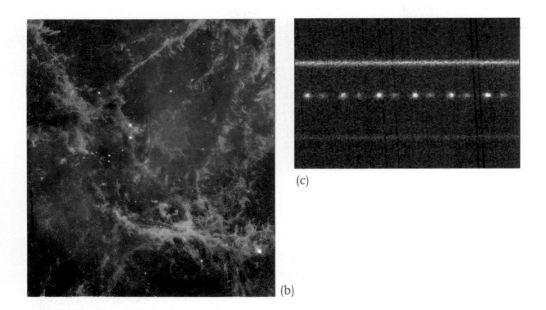

(c)

(b)

the appearance of such a new star in the constellation Taurus. It was six times as bright as the planet Venus, usually the brightest object in the night sky, and its position coincided with the one where today we see the Crab nebula. The Crab is at about 5000 light-years from us, therefore well within our galaxy, whose size you may remember is 100 000 light-years. Early last century, studies of the gas in the nebula revealed that it was expanding rapidly with a velocity of about 1500 km/s (900 miles/s) which is about 3 million miles per hour. A backward extrapolation of this – running the movie back in time – clearly showed that it must be the remnant of the new star observed by the Chinese in 1054. Curiously, there are no surviving European records of the event, maybe because nobody was interested in recording it, or possibly because it was just not supposed to happen since the stars were "fixed and immutable." The properties of the Crab nebula match what we expect from an explosion. What we see today is the remnant of a star of more than eight times the mass of our Sun, which had gone through all possible nuclear reactions and ended its life in a cataclysmic explosion: the supernova. Of the two faint stars near the center of the nebula, the southern one is quite peculiar with a very hot surface. This star, apparently the center of the explosion, held a surprise for astronomy, as we shall see below.

Iron, the most abundant element on Earth, is a special element, not only because we use it to manufacture all sorts of things, but more significantly because it establishes a turning point in the evolution of those relatively rare stars of mass more than about eight times the mass of the Sun. Such stars are short lived in cosmic terms, lasting only a few million years – a cosmic instant. Once the core of a massive star reaches, through nucleosynthesis, a stage where it is composed of iron, there are no further nuclear reactions which can *produce* energy. That's why iron is special. The massive star, now lacking a source of energy to provide the internal pressure necessary for equilibrium, collapses. In a matter of seconds all hell breaks loose. For a brief instant, as the star collapses, the temperature increases to unheard-of values, leading to a final flash of nuclear reactions. In one short second, the central one-solar-mass of material is compacted by a factor of 1 million to reach an incredibly high density. Protons and electrons in the core combine to produce neutrons generating neutrinos as a byproduct.

The neutrinos are produced in huge quantities, while the stellar matter surrounding the core becomes fantastically dense, billions of times denser than the Earth. Flying out at the speed of light, the neutrinos help push the outer layers of the dying star into space at great speeds. The star blows itself apart in a mind-boggling explosion, one of the most energetic events occurring in our Universe. In a few days the brightness of the dying star increases by a factor of 100 million, the explosion becoming visible from a great distance. The outer layers of the dying star now form a gas shell rushing out at speeds of millions of miles per hour from the center, creating an ever-expanding region containing the debris of the star. At this velocity, it would take only a few hours to travel from the Sun to the Earth. During the first seconds of the explosion, energetic neutrons and protons collide with the lighter atoms produced during the life of the star, and in a brief instant produce all the elements of the periodic table heavier than iron, including platinum, gold, and iridium. Because of the unique way in which these elements are produced, their average observed abundance is much less than that of the lighter elements. Some elements formed in this last flash are long-lived radioactive atoms such as uranium, thorium, and potassium. Their billion-year half-lives mean that they still will be around in significant quantities when some future planetary system forms, and will release their stored energy on time scales given by their half-lives, heating the planet and inducing geophysical processes. The supernova explosion returns several solar masses of material, now enriched with the heavy chemical elements, to the interstellar medium.

Perhaps you will now look at the gold and platinum in your jewelry from a different perspective. These metals are expensive because they are rare and beautiful, but what you should value is the thought that what you hold in your hand was formed more than 5 billion years ago in the explosion of a

The cosmic abundance of the elements

Fusion reactions produce heavier elements, up to iron (Fe), at higher temperatures in more massive stars. Some examples:

$$3 \times {}^4\text{He} \rightarrow {}^8\text{Be} + {}^4\text{He} \rightarrow {}^{12}\text{C} \qquad {}^{12}\text{C} + {}^4\text{He} \rightarrow {}^{16}\text{O} \qquad \text{(at 100 million °C)}$$
$$2 \times {}^{12}\text{C} \rightarrow {}^4\text{He} + {}^{20}\text{Ne} \qquad\qquad\qquad\qquad\qquad\;\; \text{(at 600 million °C)}$$
$$2 \times {}^{16}\text{O} \rightarrow {}^4\text{He} + {}^{28}\text{Si} \qquad\quad 2 \times {}^{16}\text{O} \rightarrow {}^{32}\text{S} \qquad \text{(at 1500 million °C)}$$
$$2 \times {}^{28}\text{Si} \rightarrow {}^{56}\text{Fe} \qquad\qquad\qquad\qquad\qquad\qquad\;\;\; \text{(at 4000 million °C)}$$

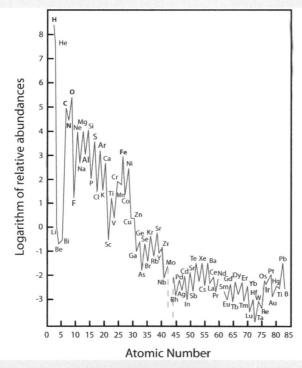

On this graph each vertical interval on the left represents a factor of 10 change. The abundances are obtained from studies of the Sun and meteorites. Ninety-eight percent of the mass is in hydrogen and helium. The zigzag pattern is a consequence of nuclear reactions which favor the formation of nuclei with even atomic number (the number of protons in the nucleus). For every 1 000 000 atoms of hydrogen there are approximately 50 000 of helium, 350 of carbon, 110 of nitrogen, 670 of oxygen, 60 of neon, 34 of magnesium, 34 of iron, 35 of silicon, and 6 of sulfur. Note the rarity of lithium, beryllium, and boron, and the relative peak about iron (Fe).

giant star somewhere in our galaxy. As you will learn in the next chapter, this material eventually became part of the Earth as it formed out of a dense interstellar cloud. Since massive stars are rare, so are supernovas, occurring at a rate of about three per century in our galaxy. Despite this rarity, over the 12 billion years of our galaxy's history these explosions have enriched the interstellar medium with an equivalent of perhaps as much as a billion solar masses of heavy elements. Without these, solar systems with Earth-like planets could not have formed.

Cosmic tombstones

All astronomical knowledge from the time of Galileo's first observations with his small telescope until the 1930s was obtained with an ever-improving set of optical telescopes. There was, however, an entire aspect of the Universe hidden from our view. In the late 1940s and early 1950s, advances in the new science of radio astronomy opened a new window to the Universe. For the first time, humans were able to study the *invisible* Universe, discovering objects which sometimes had no visible counterpart, and finding that what seemed to be ordinary stars were instead the most distant objects in the Universe, named quasars. The cosmic microwave background radiation, discovered with a radio telescope, is among the great discoveries of the previous or any century. It is the remnant of the biggest explosion ever, the one that started the whole show and is appropriately known as the Big Bang. Studying the "21-centimeter line" – radio waves emitted by the atom of hydrogen at a wavelength of 21 centimeters – radio astronomers have measured the masses of galaxies, and have contributed to studies of the structure of the Universe. All these wonderful discoveries might be the subject of another book, so here I note only that the Crab nebula is also one of the brightest radio sources in the sky. It was the first object detected with a radio telescope for which an optical identification was established, and it had a further big surprise in store.

Late in 1967, Jocellyn Bell, then a graduate student working with Anthony Hewish in Cambridge, England, discovered something unexpected and extraordinary. Using a new radio telescope built specially to study the properties of distant quasars, she noticed rapid radio pulses arriving from a specific direction in the sky. It was soon realized that these extremely regular pulses of radio radiation were not signals from a distant civilization – probably to the disappointment of some – but instead related to some previously unknown natural phenomenon, soon christened as "pulsars." Theoretical work quickly led to the conclusion that pulsars had to be

The largest telescope on Earth is found near the town of Arecibo on the green island of Puerto Rico. The big "dish" (the reflector), suspended from cables over a large depression in the ground – a sinkhole in the Karst and not an extinct volcano as is sometimes stated – measures 1000 feet (305 meters) in diameter. Faint radio waves produced by interstellar gas in distant galaxies or pulses originating in pulsars of the Milky Way are captured by the giant reflector, focused inside the suspended dome, and then transformed into electrical signals that are analyzed by astronomers. The instrument is also used as a powerful radar to study the properties of planets, their moons, and smaller objects in our solar system such as asteroids and comets, especially those which come close to the Earth. It is a large "eye" which peers into the Universe capturing the faint signals that bring to us news from faraway places. Maybe, one day, it will also bring us news from faraway beings. (NAIC Arecibo Observatory, a facility of the NSF. Photo by the author)

neutron stars, objects until then considered more appropriate for a science fiction story. A neutron star is obtained by compressing matter to such a high degree that it becomes as dense as the matter in the nuclei of atoms. If we were to compress the Earth to reach this density it would fit into a football stadium. Or think about this: if you were to squeeze the population of the Earth – 6 billion people – into a small can of sardines, it would weigh as much as if you had filled it with neutron star material. The place where this can in fact happen is in the compressed stellar core produced by a supernova during the final moments of the explosion. A neutron star is small, only about 10 miles in diameter, but has as much mass as the Sun. As you may remember from the last chapter, this is about 330 000 times the mass of the Earth. A neutron star will spin rapidly because of the contraction of the spinning core of the star, in the same manner that an ice-skater spins more rapidly as she contracts her arms.

Confirmation of the idea that pulsars are born out of the ashes of a super-nova came in 1968 when a pulsar spinning at an astonishing rate of 30 times per second, faster than your kitchen blender, was discovered at the center of the Crab nebula. It was identified with the strange faint southern star at the center of the nebula, showing that this must indeed have been the center of the explosion. This established the relationship between supernovae and neutron stars, which present themselves to us in the guise of pulsars. More than 1000 pulsars have been discovered since 1968, beacons in space which, like cosmic tombstones, signal the location of the remains of once majestic stars.

In reality pulsars do not pulse. These spherical flywheels emit two highly directional beams of radio waves in opposite directions along their magnetic axis, produced by electrons accelerated in this very strong magnetic field. Because this axis does not coincide with the rotation axis these beams sweep around the sky once per stellar rotation, much like the beacon of a light house. An observer receives a short pulse of radio waves each time one of the radio beams points at the Earth. The regularity in the period is phenome-nal: astronomers can predict the arrival times of pulses ("ticks from the pulsar") a year ahead with an accuracy better than a thousandth of a second. Some of the vast reservoir of rotational kinetic energy is converted (by a mechanism that remains unknown) into radio pulses and therefore the neutron star must be slowing down. As a pulsar radiates away its rotational energy, its pulse period gradually increases, and the energy in its emission beams decreases, until, after about 10 million years, it disappears from the radio sky.

The 1974 Nobel Prize in Physics was awarded to Anthony Hewish (1924–) for his role in the discovery of pulsars. He shared this prize with Sir Martin Ryle (1918–1984), who was awarded the prize for his pioneering research in radio astrophysics.

Supernova 1987A

One hundred and sixty thousand years ago, a massive star called Sanduleak –69˚202 ended its life in a cataclysmic explosion. This was observed as super-nova 1987A, the first supernova in 1987. Located in the Large Magellanic cloud, it was discovered on February 24, 1987, by Ian Shelton, a Canadian astronomer, right after developing a photographic plate he had just taken, using a telescope high in the Andes mountains in Chile. As he walked out to look at the Large Magellanic cloud, he could clearly see the new star. It must have been an awesome moment, witnessing the process of destruction of a star, and in some manner, simultaneously, that of creation. We know that the light of the explosion arrived on Earth between 9:30 and 10:30 on February 23 because at 9:30, Albert Jones, an amateur astronomer, saw nothing unusual when observing the Large Magellanic cloud, whereas at 10:30 the supernova was photographed.

The two Magellanic clouds, the large and the small one, visible with the naked eye from the southern hemisphere, are small companion galaxies of the Milky Way, found about 160 000 light-years from Earth. Recall that it takes only 8 minutes for light to arrive from the Sun, while the light we see

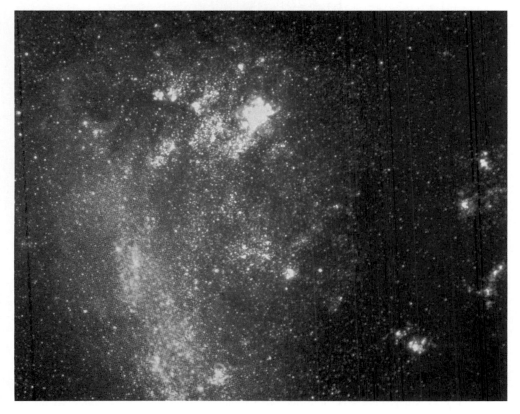

The Large Magellanic cloud (LMC) is the larger of two small companion galaxies to the Milky Way. They are the nearest galaxies to us. They each contain "only" several billion stars, so that these systems are small only when compared with large galaxies, like our Milky Way or Andromeda, which contain 100 times as many stars. The reddish Tarantula nebula, site of supernova 1987A, can be seen toward the upper right of the LMC. This photograph was obtained with the Cerro Tololo Interamerican Observatory's 4-meter Blanco telescope located on the Chilean Andes. (AURA/NOAO/NSF)

Supernova 1987A was the first supernova observed in 1987 (hence the A). It was discovered on February 24, 1987, when a star, until then inconspicuous, brightened to the point that it became visible to the naked eye, even at its enormous distance of 160 000 light-years. It is the bright image at the lower left of this frame near the red cloud of the Tarantula nebula. Compare this with the previous illustration where it is absent. The flash of light that made this image left the Magellanic Clouds on its long voyage to Earth 160 000 years ago. While it traveled toward Earth, a hominid descendant of early apes embarked on an evolutionary journey. He began to walk upright and his descendants with a larger brain, developed tools, language, and finally a technology, including telescopes, in time to catch the flash as it arrived here. (E.O. Lawrence Berkeley National Laboratory)

from the Magellanic clouds left there 160 000 years ago at a time when on the African savannah an ape descendant, *Homo sapiens*, made its first appearance.

Supernova 1987A delivered a pleasant surprise for us, detected by an unlikely source. The Kamiokande II neutrino detector lies 3300 feet underground in a zinc mine belonging to the Kamioka Mining and Smelting company, in Japan. This detector contains a large tank with a few thousand tons of specially purified water, monitored by a large number of sensitive light detectors. A similar detector, half way around the Earth, belonging to the IMB (Irvine–Michigan–Brookhaven) collaboration is found in the

The Hubble Space Telescope took this detailed image of a small area of the LMC centered on supernova 1987A. Several massive bright blue stars surrounded by clouds of gas are visible near the remnant of the supernova, stars belonging to the same generation as the one that exploded. Another one could become a supernova soon, say within the next million years. The area is filled with wisps of gas out of which new stars are being formed. The remnant of the supernova is the bright dot at the center, surrounded by an inner and outer ring of bright red gas believed to be material expelled from the progenitor star many thousands of years before the explosion. In a few years the rapidly expanding gas of the supernova will overtake the bright inner ring. The collision will heat and excite the gas, producing visible effects which astronomers hope to study in the near future. (Hubble Heritage Team (AURA/STScI/NASA))

Fairport Harbor salt mine of the Morton–Thiokol corporation, 2000 feet underground near the shore of Lake Erie. These instruments, located deep underground to shield them from confusing radiation, had been built by scientists to search for neutrinos produced by the possible decay of protons (they did not find any). On February 23, 1987, at 7 hours and 36 minutes Greenwich time, both detectors experienced a unique burst of neutrino detections. Because neutrinos are so unlikely to interact with matter, the handful detected meant that about 50 billion of them must have passed each square inch of the detectors during the burst. The burst of neutrinos was produced in the core of supernova 1987A, just three hours before the observed explosion. They were detected after traveling for 160 000 years. These observations led to the amazing conclusion that the supernova must have converted the equivalent of 30 000 Earth masses into energy, most of it into neutrinos. The detection of neutrinos from supernova 1987A was a significant result for modern observational astronomy. It allowed scientists to "see," for the first time ever, into the core of the supernova's precursor star and confirm the theory of supernova explosions and the formation of neutron stars. This was also the first occasion on which neutrinos from outside the solar system had been detected.

Tycho and Kepler

Before supernova 1987A, the last supernova visible to the naked eye had been observed in 1604 by the great German astronomer and mathematician, Johannes Kepler (1571–1630). Before this, in 1572, a new star had appeared in the constellation Cassiopeia. Tycho Brahe (1546–1601), the famous Danish astronomer, whose careful measurements of the positions of stars and planets allowed Kepler to discover the laws of planetary motion, observed this supernova in November of that year. Tycho, who knew the skies as nobody else did at that time, could scarcely believe his eyes and tells us:

> Amazed, and as if astonished and stupefied, I stood still, gazing for a certain length of time with my eyes fixed intently upon it and noticing that same star placed close to the stars which antiquity attributed to Cassiopeia. When I had satisfied myself that no star of that kind had ever shone forth before, I was led into such perplexity by the unbelievability of the thing that I began to doubt the faith of my own eyes, and so, turning to the servants who were accompanying me, I asked them whether they too could see a certain extremely bright star when I pointed out the place directly overhead.[2]

[2] Quoted in David H. Clark and F. Richard Stephenson, *The Historical Supernovae*, Pergamon Press, 1977, p. 174.

The reason for his astonishment and disbelief, so eloquently narrated, was that the observation of a new star ran counter to the established Aristotelian view that the eternal eighth sphere of the heavens, containing the fixed stars, was perfect and immutable. In fact, this event observed by Tycho was one of several which contributed to the great intellectual change known as the Copernican Revolution.

Kepler, a contemporary of Galileo, was born in 1571 in Weil-der-Stadt, near Stuttgart in Germany. He grew up in a poor household, an unhappy, sickly child raised by his grandparents, quite a contrast to the noble and wealthy Brahe. He attended the University of Tübingen and there learned astronomy from Michael Maestlin, one of the leading astronomers of the time. In 1597 he began to work at a school in Graz, Austria, as a teacher of mathematics and astronomy. In the year 1600 he accepted an offer to work at Brahe's new observatory near Prague as his assistant. A year later, upon Brahe's death, he was appointed his successor as imperial mathematician to the court of Emperor Rudolph II. He applied himself to the study of the motion of Mars, based on the precise measurements of Tycho but, try as he might, he could not describe the measurements as a result of circular motion at constant velocity. It took many years before he abandoned the idea that it had to be circular motion. In his *Astronomia Nova,* published in 1609, he showed that the orbit of Mars was an ellipse, with the Sun occupying one of its two foci. This represented a radical departure from the past, since until then it was supposed that "perfect" celestial objects should travel in "perfect" circular orbits. In 1619 Kepler published *Harmonice Mundi,* completing his work on the principles of planetary motions. Now, with elliptical orbits and without the need for cumbersome epicycles, the heliocentric system worked like a charm. The road was now clear for Isaac Newton to explain these elliptical orbits as a consequence of his law of gravitation. After a difficult life, Kepler died on November 15, 1630.

Fertilization of space

Evolution of a star depends primarily on its mass because, as we have seen, this determines both the amount of "fuel" available and the maximum central temperature attainable for the different possible fusion reactions to proceed. During the course of their life, stars will slowly lose mass into space. For example, the Sun releases protons and electrons generating what is called the solar wind. This wind of energetic particles causes the beautiful auroras, the northern and southern lights, the result of solar flares which intensify the solar wind. Did you know that comet tails always point away

This is how the northern sky looked on November 28, 2000, over the port city of Nome, on the west coast of Alaska, when a solar storm overtook the Earth and painted the night sky. This impressive display of auroras is produced when particles with high energies composing the solar wind strike the atmosphere, ionizing oxygen and nitrogen atoms in our atmosphere, which then glow with different colors. (John Russell (northernlightsnome. homestead.com))

from the Sun, no matter which direction the comet is moving in? This happens because the material in the tail is swept away by the outflowing solar wind, just as when you blow away the smoke from a candle.

When a star gets old, it becomes larger, cooler, and redder, increasing its energy output tremendously. It leaves the hydrogen fusion stage and becomes a red giant. If its mass is less than about eight times the mass of the Sun, it does not explode as a supernova. The star begins generating a much stronger wind, perhaps 1 million times stronger than before. Over several million years, some giant stars can, in this way, lose a substantial fraction of their mass, slowly expelling their surface layers and contracting their cores. The gas that leaves the surface travels into cold space and condenses into microscopic interstellar grains composed of carbon and silicates. Millions of years later the grains will collect into dark interstellar clouds, mixing with the material produced by supernovae. Out of this fertile raw material new stellar systems will form, as we shall see in the next chapter.

Because not enough mass is available to increase the central temperatures in these stars, the generation of energy by fusion in the core eventually ceases, and in time they settle to become what are called white dwarf stars. They are white because they are still very hot, some reaching surface temperatures of 100 000 °C, about fifteen times hotter than the Sun, and dwarfs because they are small for a star, about the size of the Earth. At such high temperatures these small stellar remnants are not very bright in visible light because most of their radiation is emitted as invisible ultraviolet radiation. These stellar cinders, with the mass of the Sun, are exotic objects, not as weird as neutron stars, but still made of very dense material. Electrons inside a white dwarf star are squeezed so tightly that the object cannot contract further, although the force of gravity at its surface would make a 130-pound human weigh one million pounds. Clearly not a place to visit.

The intense ultraviolet radiation from the central white dwarf will ionize the surrounding shell of expelled material, and make it glow. The glowing shell, leisurely expanding at a velocity that is only one-hundredth of that of the expanding debris of a supernova, is called a planetary nebula because of

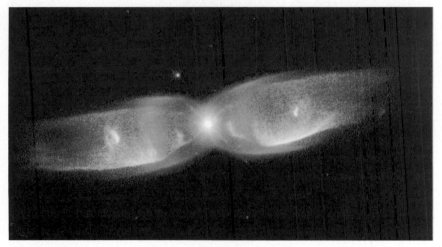

One thousand years ago, a pair of stars 2100 light-years away in the constellation Ophiucus got too close to each other, disrupting equilibrium. A high-speed stellar gas wind in two opposite jets was the consequence, ejected from the surface of one star by the gravitational pull of the other. The result is this strikingly beautiful cosmic display called the M2-9 nebula. In this image obtained by the Hubble Space Telescope in 1997, the colors are produced by the light emitted by different atoms. (Bruce Balick (University of Washington), Vincent Icke (Leiden University, The Netherlands), Garrelt Mellema (Stockholm University), and STScI/NASA)

its disk-shaped appearance when observed through a small telescope. However, it has nothing to do with planets. The planetary nebulae are among the most beautiful objects in the sky, as you can appreciate in the photographs.

By this process the elements produced inside the star will be dispersed into the interstellar medium, ready to be incorporated into new stars and planetary systems. However those elements incorporated into the white dwarf will be forever locked away and will no longer be recycled. A significant amount of the carbon and particulate matter in interstellar space is manufactured and dispersed by red giant stars in this manner. Although not as spectacular as the supernova, this gentle death of stars contributes a few solar masses of recycled material to the interstellar medium each year, and over the history of our galaxy generates about 12 billion solar masses of chemical elements, which enrich the interstellar medium. About 5 billion

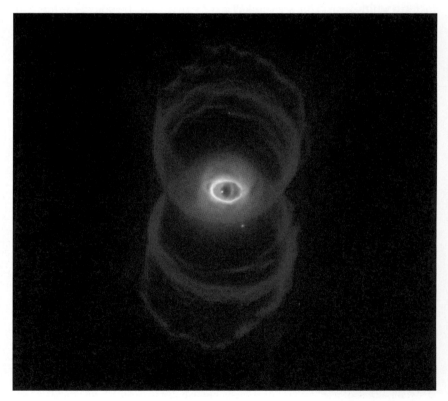

The hourglass planetary nebula, is located about 8000 light-years from Earth. The red gas is nitrogen whereas towards the center the glow is dominated by light from hydrogen (green) and oxygen (blue). The shape of the nebula is in the form of an hourglass, formed by successive episodes of mass ejection form the central star. The slowly expanding gas, containing atoms produced by the central star, will slowly dissipate until the material joins smoothly with the interstellar medium. The faint central hot white dwarf star will glow for a long time after the nebula dissipates. (STScI/NASA)

years from now, the nebula surrounding our white dwarf Sun, which in the previous chapter had already swallowed Earth, will slowly dissipate into space until no trace of it will be left.

Understanding the energy generation process in stars, and their evolution, is one of the great achievements of modern astrophysics, but what is of interest in the context of this story is that it involves the transmutation of lighter elements into heavier ones, a kind of cosmic alchemy. If you imagine a Universe where all you had initially was the simplest element, hydrogen, you can see how through the process of nucleosynthesis all the chemical elements of the periodic table could be obtained. So, as time passed, and several generations of massive stars came and went, the interstellar medium could become enriched with carbon, oxygen, iron, and all the heavier elements needed to form Earth-like planets. This is in fact what happened.

Children of the stars

In the 1920s the American astronomer Edwin Hubble (1889–1953) established that galaxies, then known as nebulae, were large independent stellar systems at enormous distances from us, and not components of our own Milky Way galaxy. Our galaxy became just one of billions of similar systems and our place in the Universe became even more insignificant than before. Hubble also discovered the fact that all galaxies are moving away from each other, the "expansion" of the Universe, one of the pillars of modern cosmology. This led to the idea that at some time in the very distant past all the matter in the Universe was concentrated to an almost infinite density, in an infinitesimal volume, at an extremely high temperature: the Big Bang. Theoretical research predicted that as the Universe expanded from this initial state, the radiation it contained would cool so that today, after some 15 billion years of expansion, it would be at a temperature of only a few degrees above absolute zero, the lowest possible temperature. Theory also predicted the precise shape of the graph describing this radiation, a shape which was the same as that obtained by laboratory measurements of hot objects. It is called "blackbody" radiation.

In 1965, German-born Arno Penzias (1933–) and Robert Wilson (1936–) were tuning a very sensitive communications antenna belonging to the Bell Laboratories, in Holmdel, New Jersey. In the course of their work they noticed that their antenna was receiving radiation which seemed to come from nowhere in particular, that is, it was everywhere. Pursuing the source of this excess radiation like a pair of detectives on a hot trail, they were led to the conclusion that what they were detecting was the predicted afterglow of the Big Bang. NASA's Cosmic Background Explorer satellite (COBE), launched into Earth orbit in 1989, spent several years studying the properties of this cosmic afterglow and established, with unprecedented accuracy, the fact that the shape of the graph describing this radiation is exactly that of a blackbody at a temperature of only 2.7 °C above absolute zero. This magnificent discovery represents the second pillar of modern cosmology.

It is astonishing that from our insignificant vantage point we can say anything about events that happened 15 billion years ago, at a time when the Universe was a hot unstructured place without galaxies, stars, or planets. The fact that the speed of light is very high but not infinite allows us to study the history of the Universe and its birth – actual genesis. Thus, when we look at a distant galaxy, we look into the past seeing it as it was millions or billions of years ago, when the light of its stars left on its long journey to our telescopes. Our telescopes are in this way converted into time travel machines. The cosmic afterglow comes to us from a time only a few hundred thousand years after the Big Bang, a blink of the eye in cosmic history.

The 1978 Nobel Prize in Physics was awarded to Penzias and Wilson for their discovery of the Cosmic Microwave Background Radiation. While on the topic, I mention that the 1983 Nobel Prize in Physics was awarded to Indian-born Subrahmanyan Chandrasekhar (1910–1994), for his important contributions to the theories of stellar structure and evolution, and to William Fowler (1911–1995) for his contributions to the study of nucleosynthesis in stars.

Modern studies of nucleosynthesis attempt to predict all the hundreds of possible nuclear reactions occurring in stars at different stages in their evolution, and in the aftermath of a supernova explosion. There is remarkable agreement between these calculations and the observed abundance of the elements. These studies have provided the answer to the profound question of the origin of the elements, one of the greatest discoveries of the last century of the second millennium. The oxygen and nitrogen we breathe, the aluminum and other metals in our airplanes, the gold and platinum in our rings, and the carbon in our bodies were all made in stellar processes. Without stars we and our world would not exist and, without the steady source of energy from our Sun, life could not have developed on our planet. *We are children of the stars.*

All over our galaxy, as we look through telescopes, we see regions containing clusters of young stars and gas. These are regions of star formation. They must have formed recently because we still see there the short-lived massive luminous stars which last for only about 10 million years or less, sometimes shining through the dark clouds of gas and dust out of which they formed. We shall see in the next chapter how this happened.

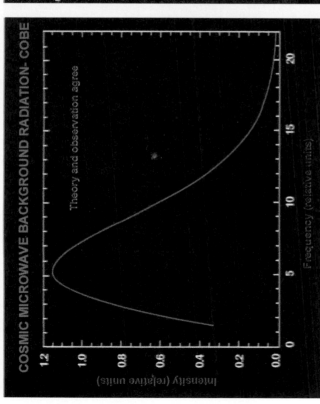

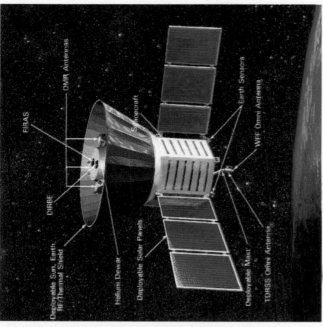

On this graph the continuous curve is the theoretically predicted spectrum (the intensity of radiation as a function of the wavelength) of a blackbody radiator at a temperature of 2.73 degrees above absolute zero. Measurements were obtained by the Far Infrared Absolute Spectrophotometer (FIRAS) on board of NASA's COBE (Cosmic Background Explorer) satellite, which operated for more than four years. The agreement between the predictions of the Big Bang theory and the data is so good that the experimental points cannot be distinguished from the calculated smooth curve. This is remarkable and represents the best proof of the Big Bang theory. The illustration on the right shows the spacecraft. It is roughly the size of a small truck, weighing about 4500 pounds (2000 kg), and was launched in November of 1989. The COBE instruments were pointed away from the Earth and a protective shield prevented any sunlight or light reflected from or emitted by the Earth from entering the instruments. This gave them the best possible view (from Earth orbit) of the Universe. (NASA Goddard Space Flight Center and the COBE Science Working Group)

The planets were born out of the violence and heat from countless collisions between objects in the disk of material surrounding the young Sun some 4 billion years ago, as illustrated in this painting.

Chapter 3

The birth of planets

Made in Magrathea

The Universe was ready to build planets where life could arise only after it had produced all the necessary elements needed for this to happen. We have sent robots to explore most of the planets of the solar system and have personally visited the Moon. These studies have allowed us to understand the huge difference between these worlds.

Dark clouds

Somewhere near one of the spiral arms of our Milky Way galaxy, thousands of light-years away from us, in the cold and dark confines of interstellar space, a giant shock wave produced by a distant supernova explosion gathered up interstellar material. It did this like a snowplow gathers snow, just taking millions of times longer, and formed a gigantic tenuous cloud of material. This giant interstellar molecular cloud, slowly contracting under its own gravity, became denser and eventually, after millions of years passed, blocked out the light from background stars. Today, it appears as a dark cloud, as if it were a hole in the sky. This is truly a giant cloud, so large that light takes tens or even hundreds of years to travel from one end to the other. The cloud is composed almost entirely of hydrogen and helium, the two simplest elements, as is most of the rest of the Universe we observe.

When we aim sensitive radio telescopes at such a cloud, we discover that there are also traces of other compounds such as carbon monoxide (CO), formaldehyde (H_2CO), methanol (CH_3OH), water (H_2O), and hydrocyanic acid (HCN). We also find a rich variety of other more complex molecules, mostly made of carbon atoms linked to hydrogen, oxygen, nitrogen and sulfur atoms, the most abundantly available elements. This is why it is called a molecular cloud. This material is there because of the many stars which during their lifetimes produced these elements and in death returned them to the

This "hole in the sky," toward the southern constellation Ophiuchus, is named Barnard 68 after the American astronomer, Edward E. Barnard (1857–1923), who included it in a list of such objects. This dark cloud of dust and gas is sharply defined against a rich background of stars in the Milky Way. Not one foreground star is observed because the cloud is relatively nearby, at a distance of 500 light-years. It is about one light-year in extent. (ESO, FORS (Focal Reducer low-dispersion Spectograph) Team, VLT Antu)

interstellar medium. These organic molecules can be quite complex, as in the eight-atom molecule of the sugar glycoaldehyde ($C_2H_4O_2$), so it is surprising that they can survive the harsh conditions of interstellar space, which is flooded by ultraviolet radiation from many stars that could destroy them. But the cloud also contains interstellar grains, the size of smoke particles, composed of silicates – minerals containing silicon, oxygen, magnesium, iron and other common atoms – and carbon in various forms. These grains condensed out of the solar winds of red giant stars toward the end of their lives. In these very cold regions of space, ices of volatile compounds such as water and methane – substances that boil away easily – condense and form an icy coating on the surface of these tiny grains. Even more complex molecules can form on the surface of these grains, some of them similar to those found in living things. The grains shield the inside regions of the cloud from ultraviolet starlight, protecting these molecules from being destroyed. The density of gas is so low that it would be considered a very good vacuum on Earth, in fact the best vacuum chambers barely reach this level of emptiness, but it is still over 1000 times denser than the normal medium between the stars, which for all practical purposes is empty. In spite of it being almost a vacuum, the region is so large that the total amount of material is also very large. Some clouds can contain 1 million solar masses of material, equivalent to 300 000 million times the mass of the Earth. Therefore, even if these trace molecules observed by radio telescopes represent only a minute fraction of the total material in the cloud, a trace of about 100 thousandths of the total, the cloud still contains 10 solar masses of this material. This is enough to build 3 million planets like Earth, although this does not necessarily happen.

As the dark and very cold cloud contracts due to its own gravitation, it fragments into many smaller clouds, which continue collapsing at an accelerated pace. The density and temperature of these clouds slowly increase as they collapse, and any initial small rotation which they might have had is amplified, in the same way that an ice-skater spins faster when he contracts his arms. After a few million years, a short time for cosmic phenomena, each cloud has reached a point where at the center, heated by the in-falling material, a self-luminous blob appears: *a protostar is born*. In some clouds, part of the remaining material will settle into a rotating disk surrounding this protostar, a *protoplanetary disk* which contains the material out of which planets will form.

If you were to look at all this from very far away and be able to speed up the action by a factor of several millions, it would look like Fourth-of-July fireworks. The dark sky would light up with several hundred points of light, not all at the same time and of the same color. Some, the brightest, would maybe last only about a year and then with a flash of brilliance disappear, while others, dimmer and redder, would last for a long time. The thousands of stars in the Pleiades star cluster, about 400 light-years from us, are such a fireworks, frozen in time since we are not really able to speed up the action. How many stars in the Pleiades you can see with the naked eye will depend on your eyesight, but most people can see seven, and so the cluster is also known as the "seven sisters" and, if you look at it with binoculars, you will see the beautiful display of hundreds of stars in the cluster.

Astronomers would like to understand the detailed processes that led to the formation of the planets and resulted in the variety we observe, from the inner rocky planets to the outer gaseous giants. This composite of images obtained by various spacecraft shows the four inner rocky Earth-like planets scaled to the correct relative sizes, and the four gaseous giants, which are however shown smaller than they should be relative to the Earth-like ones. Pluto is not shown, not because it has been demoted from planethood, but because it has yet to be visited by a spacecraft. (NASA/JPL/Caltech)

The Pleiades star cluster contains more than 3000 stars, but only six to twelve of the brightest ones can be seen with the naked eye. These are surrounded by intricate blue wisps of light, the result of starlight reflecting off minute grains of interstellar dust in the vicinity. (Main image Matt BenDaniel; smaller image J. Salgado and C. Vaughn (www.aa6g.org))

This photo of Orion shows you its main features. The bright stars forming this constellation are not physically related, as are the stars of the Pleiades. They are at different distances. Betelgeuse, the brightest star at the top left is a red supergiant star about 500 light-years away while bluish Rigel, at bottom right, is 1000 light-years away. Betelgeuse is cooler than the Sun, more massive, and more than 1000 times larger. It is nearing the end of its life and will become a supernova sometime soon, in cosmic terms. The great nebula (M42) is visible as the middle "star" of Orion's sword. If you look at it with binoculars you can see that it is not a star. The faint loop of red luminosity is the emission nebula called Barnard's Loop. (Till Credner, AlltheSky.com)

Even if you are not familiar with the constellations, you will recognize Orion, the mighty hunter and warrior, because it is strikingly easy to find in the northern winter skies. Look at the accompanying picture and you will be able to identify it. It is formed by the bright red supergiant Betelgeuse at the top left as you look up, the less-bright pale yellow Bellatrix to the right, bright bluish Rigel at the bottom right, and Saiph, not as bright but still quite prominent, at the bottom left, forming a not quite rectangular figure. Diagonally across the middle of this rectangle are three bright stars, Mintaka, Alnilam, and Alnitak, which line up to form Orion's belt. From the belt hangs Orion's sword, formed by three fainter stars. If you look at the middle star of this sword with binoculars you will discover that it is not a

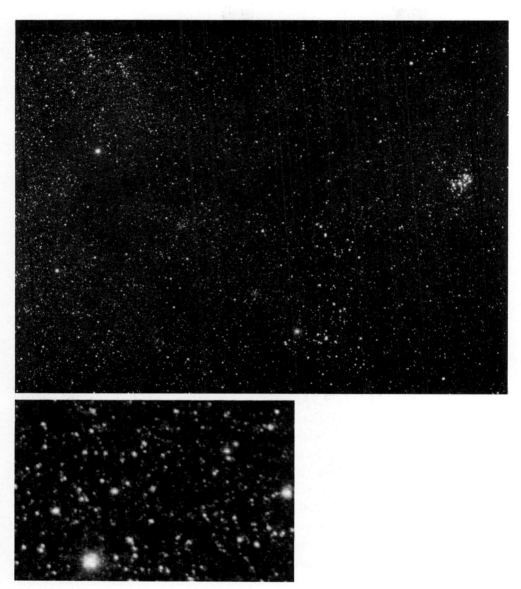

At a distance of about 400 light-years, the stars of the Pleiades cluster (M45) shine like
jewels in the northern sky. The bright orange star at bottom center is Aldebaran ("the
follower" – of the Pleiades), the thirteenth brightest star in the sky, at a distance of 61
light-years. To the south of it (bottom) is a very loose cluster of yellow stars, the Hyades
cluster, 150 light-years from us. The bright star one-third up and near the left edge is
Zeta Tauri about 500 light-years away. You can see its surroundings on the blow-up, at
the center of which you can see a small irregular reddish smudge. *That* is the Crab
nebula. Orion lies out of this view to the bottom left. Some of the stars on the bottom
left can also be seen in the photo of Orion. Can you recognize them? (The photos are not
to the same scale.) (Till Credner, AlltheSky.com)

star at all but instead a fuzzy object, in fact the forty-second fuzzy object in Messier's catalog (M42), called the Orion nebula. Not far away, about 14 degrees to the north of Betelgeuse, in Taurus constellation, lies the Crab nebula which we met in the last chapter, and if you look about 30 degrees to the west and a bit to the north from this you will find the Pleiades. Try it.

The clearest and most detailed views of the Orion nebula, a relatively small region at the edge of a giant molecular cloud which is at a distance of 1600 light-years from us, have been obtained with the Hubble Space Telescope. This marvelous instrument, which is above the Earth's atmosphere, can obtain pictures of distant objects with astonishing detail. Inside the Orion nebula we find thousands of bright, recently formed stars, stellar nurseries, where "recently" in cosmic terms means several million years.

Planet birth

About 5 billion years ago, the cloud of interstellar gas and dust that formed the solar system, collapsed, and the central dense region contracted to form the proto-Sun. A rotating disk of material surrounded it: the solar nebula. This disk extended to a very large distance, tens of times farther than the orbit of Pluto, the yet-to-be-born outermost planet of the solar system. Moving away from the young Sun, the material of the nebula was more rarefied, and the temperature, which was quite high near the center, decreased. This established in great measure the final chemical composition of the planets that formed out of the nebular material. As the minute ice-covered interstellar grains, now in the denser parts of the solar nebula, orbited the Sun, they collided with each other and stuck together to form larger pebble-sized particles. These grew further until they were objects a mile in diameter which could start attracting more mass from the surrounding nebula by their own gravity and continue to grow larger to sizes of several miles. These so-called planetesimals collided with each other and, depending on the velocity of the collision, either blew themselves to pieces or coalesced into still larger bodies. In the end, several hundred Moon-sized objects were formed which collided further until planet-sized objects emerged. The nebula was less dense farther away from the center, and material moved more slowly farther out, in accordance with the laws discovered by Kepler. Therefore, far from the Sun, it took a long time for objects to collide and grow into planet-sized bodies, and after some distance, beyond the orbit of Pluto, this never happened.

Near the proto-Sun, the temperature was very high so that water and other volatile compounds did not form or, if present, evaporated from the

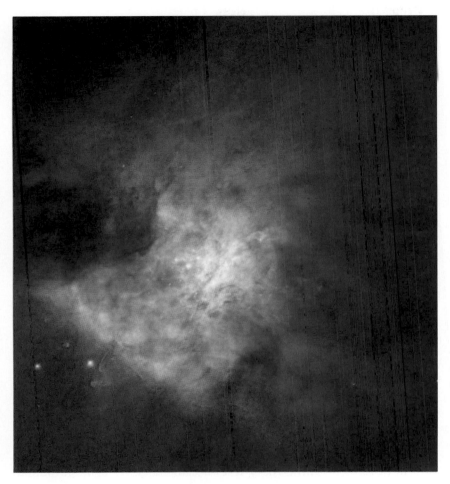

The great nebula in Orion (M42) as photographed by the Hubble Space Telescope. The image covers the inner 2.5 light-years of this large nebula, an immense, relatively nearby region, about 1600 light-years away. Besides housing a bright open cluster of recently formed stars known as the Trapezium, seen near the center of the image, the Orion nebula contains many stellar nurseries. These nurseries contain hydrogen gas, hot young stars, protoplanetary disks, and stellar jets spewing material at high speeds. Most of the filamentary structures visible are shock waves – fronts where fast-moving material encounters slow-moving gas. Shocks are particularly apparent near the bright stars in the lower left of the picture. (C. O'Dell and S. Wong (Rice University), STScI/NASA)

interstellar grains. The objects in this region of the nebula formed out of silicates, becoming the terrestrial planets: Mercury, Venus, Earth and Mars. They formed quite rapidly, say in several million years. Because they were close to the young Sun, stellar winds, which are quite strong in young stars, rapidly began to push material from the nebula away from their surroundings.

The cluster of stars NGC2244 is located in the Rosette nebula 5200 light-years from us in the Milky Way. It contains several very bright and massive stars seen at bottom center, some over twenty times the mass of the Sun. These stars have a very short lifetime, and a supernova will explode in the not too distant future. The strong stellar winds produced by these stars are blowing the surrounding gas and dust away and, as the material gets compressed, this will trigger new star formation. The red glow is produced by hydrogen atoms. (J.-C. Cuillandre and G. Fahlman CFHT)

Farther out, water ice was stable and the giants Jupiter and Saturn formed. Once their central rocky cores became large enough, gravitation could attract gas from the surrounding nebula before it was blown away by the solar wind. The final product consists of an Earth-like core surrounded by a giant mass of hydrogen and helium not very different in composition from that of the nebula and the Sun. Still farther out, at lower temperatures and densities, other ices formed, made of the most abundant elements available, such as those of the important simple compounds ammonia (NH_3), methane (CH_4), and carbon dioxide (CO_2). Whereas water freezes at 32 °F (0 °C) ammonia does so at −27 °F (−33 °C), carbon dioxide at −70 °F (−57 °C) and methane at a very cold −296 °F (−182 °C). Compounds of ammonia are used as a fertilizer, to produce sources of nitrogen for plants, and in solution ammonia is a very good solvent, which is why it is used in cleaning products. Methane is a flammable gas, the main component of natural gas. Bacteria in our guts produce it as a byproduct of the digestive process. Carbon dioxide is the most important source of carbon for living systems and controls the surface temperature of a planet. It is used to put the "fizz" into carbonated drinks and, when frozen, it is called "dry ice." All these are important compounds.

Distant planetesimals were composed of these ices and small amounts of silicates. Uranus and Neptune formed here, but they took ten times longer to grow because of the lower density in this region of the nebula. By the time their rock and ice core had grown sufficiently, the intensifying solar wind blew away the nebula, and so they did not grow as much, remaining quite a bit smaller than Jupiter and Saturn.

Beyond Neptune, the planetesimals never managed to collide and grow into large planets. True, there is Pluto, the smallest planet of the solar system, with a diameter of just 1400 miles (2250 km), only a bit over two times as large as the largest asteroid, Ceres, and with a mass less than one-fifth that of our Moon. Pluto travels in an orbit which is more highly inclined and eccentric than that of any other planet, so eccentric that for 20 years of its 248-year-long orbit about the Sun it is closer to it than Neptune is. It has even been suggested that Pluto is not really a planet at all but a large remnant left over from the time when the solar system formed. It might just be the largest of hundreds of objects which have recently been discovered in orbits beyond Pluto in the so-called Kuiper belt.

Gerald Kuiper (1905–1973) was a Dutch-American astronomer who in 1951 postulated the existence of a disk containing billions of leftover planetesimals surrounding the Sun. The Kuiper belt extends to about 500 AU, which on the scale of the model described in the first chapter, where the Sun was 700 feet from the Earth, corresponds to a distance of about 70 miles (110 km). Other objects the size of Pluto could be there, but being so far away, and therefore faint, they are very difficult to find. Of course, Pluto does not care what we call it but, if it did, it might prefer to be called the largest Kuiper belt object instead of the smallest planet.

Over the first half billion years, the young Sun developed a strong stellar wind which cleared the nebular gas, and the process of collisions and planet growth gradually swept clean the inner regions of the nebula. Many planetesimals were flung out of the nebula by the gravitational force imparted as they passed near one of the newly formed giant planets, and the planets changed their orbits in ways we do not know. Billions of planetesimals ended up in a giant cloud which surrounds the solar system today. This cloud is called the Oort cloud, after Jan Oort (1900–1992), a Dutch astronomer who in 1950 inferred the existence of this cloud from a study of cometary orbits. The cloud begins at the outer edge of the Kuiper belt and reaches perhaps as far as 50 000 AU. The Oort cloud surrounding the Sun is so large that, on the

These four images, taken within a few hours, show the motion of a faint object at the center, relative to the "fixed" background stars. It is the Kuiper belt object (KBO)1992QB1, the first of dozens discovered since. It was observed using a 2.2-meter telescope belonging to the University of Hawaii and located on Mauna Kea. The object has an estimated diameter of 100 miles. It is estimated that there are roughly 1 billion objects of diameter larger than 3 miles in the Kuiper belt. Pluto might just be one of the largest KBOs, and it would not be surprising if eventually other objects the size of Pluto are found. The elongated object, visible in the first three frames moving from right to left, is a main-belt asteroid which happened to be in the field of view. (David Jewitt and Jane Luu)

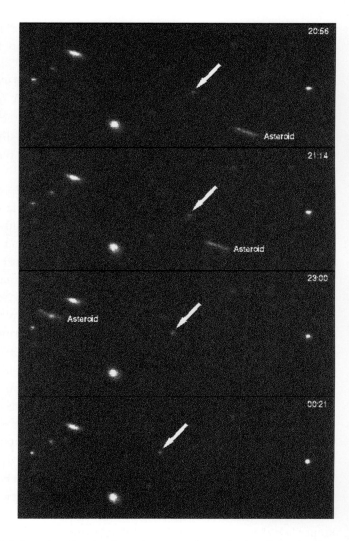

scale of the model described in the first chapter, where the Earth is the size of a penny, it would be comparable to the *actual* size of the Earth. In these frozen distant regions, the material from the solar nebula was not altered by solar heat or collisions, and the complex molecules could survive the violent processes which formed the Sun and planets. The Oort cloud and the Kuiper belt are then a great reservoir of planetesimals, a giant freezer in the solar system, preserving original material from the days of its formation. If we could only study these small distant objects close up, we would learn a lot about the epoch when the solar system formed and, as you will see, sometimes we receive a free sample.

So the formation of the planets of the solar system was a very complex process and its outcome depended on many factors. The interplay between the initial properties of the solar nebula, its temperature and density distribution, and its evolution, determined in large part by the properties of the central star, are the principal factors which influenced the outcome. This resulted in the variety we observe between the different worlds of the solar system. Many details of the process have not yet been worked out, and much is to be learned from further exploration of our solar system. Only recently have we discovered some of the inhabitants of the Kuiper belt, while the Oort cloud, being so much farther out, remains for now only a well-founded hypothesis. Take another nebula with a different mass for the central star, and things will develop in another way.

Planets elsewhere

In the last ten years it has been possible to detect planets orbiting other stars, finally answering the age-old question of whether other stars have planetary systems. Planets are too faint to be detected directly against the light of their stars, which is roughly a billion times brighter. It is like trying to see the glow of a cigarette in front of a bright searchlight. However, planets will cause the star about which they orbit to wobble slightly in space because of the gravitational force they exert on it. This can be detected as a tiny cyclic change in the measurable velocity of the star as it moves in space. In 1994, two Swiss astronomers of the Geneva Observatory, Michel Mayor and Didier Queloz, using a telescope at Haute-Provence Observatory in France, observed this small effect in the Sun-like star, 51 Pegasi. The measurements showed that a large planet with a mass close to that of Jupiter was in a very tight orbit around this star, which lies 42 light-years from us. At a distance which is almost eight times closer to 51 Pegasi than Mercury is to the Sun, this planet takes only four days to complete an orbit. Since then, the list of stars with planets has been growing, a multiplicity of strange worlds that await better instrumentation to measure their properties, and theoretical research to understand how they came to be. The study of these systems will shed much light on the details of the formation of our planetary system.

Beta Pictoris is a young Sun-like star at a distance of 50 light-years from us. It is surrounded by a disk, which we see almost edge-on, and which shines by the reflected light from the star. This disk has been observed by ground-based telescopes and by others in space. It is of great interest because it is clearly a star caught in the act of forming its planets. It's a pity that we cannot speed up time to see what Beta Pictoris will be like a few million years from now. However, several other stars have also been found with disks

Beta Pictoris, a star only 50 light-years away, is surrounded by a disk of material, viewed edge-on. It is a young Sun-like star just completing its formative stages. In this false-color image obtained in infrared light, the overwhelming light from the star itself is masked out, so that the much fainter features of the disk are revealed. Detailed studies show the inner parts of the disk to be slightly warped, possibly caused by the gravitational effect of a large planet orbiting the star, too faint to be seen directly against the stellar glare. (ESO)

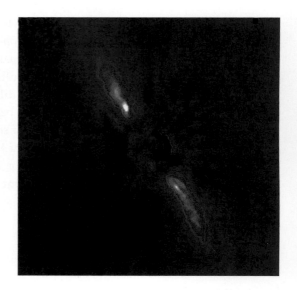

surrounding them, and the study of these, all at different stages in their development, will provide the information we need to deduce their history. It is as if you only had one second to look at a person. You would not be able to say much about the development of this individual from one image. However, if you had one second to look at 100 different persons selected at random, you would expect to be able to piece together the story of human development, going from babies to children to adults and finally old persons, and make some sense of it. This approach is often used in astronomy, to enable us to learn about the development of systems which in real-time take millions of years to change, whereas our observations only give us a snapshot, as is the case for stars. The Hubble Space Telescope obtained a beautiful snapshot of stars surrounded by large volumes of dust and gas as it pointed at the Orion nebula.

Moon and Earth

The Moon, by far the most noticeable object in the night sky, has always been of the greatest interest, and a source of fascination to humans. It is a unique satellite in the solar system, quite different from the moons of other planets, and the largest in relation to its planet (except for dubious Pluto and its moon Charon). It is large enough to cover the Sun completely in the sky. When this happens it produces one of nature's more spectacular shows – a solar eclipse. Always showing the same face toward Earth, the Moon rotates

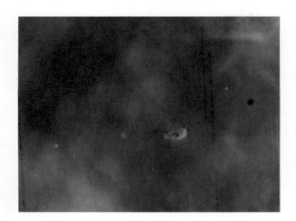

This close-up image of a region in the Orion nebula made by the Hubble Space Telescope shows five young stars surrounded by fuzzy blobs – disks of dust and gas. They are planetary systems in the process of formation. The image, which spans only about one-tenth of a light-year, shows four stars surrounded by bright nebulosity and one dark one, the black dot on the right, seen in silhouette against the bright nebula. The size of this dark blob is about ten time the size of the solar system. (C.R. O'Dell and S. Wong (Rice University), STScI/NASA)

about Earth in an almost circular orbit with a period of 29.5 days, the time between two consecutive full Moons, called the synodic period.

Other living things are also quite aware of the Moon, since the regular cycles of the tides and of nights dimly lit by the light of the full Moon (which is reflected sunlight) affect animal life. In fact many biological processes in plants and animals seem to be tied to the lunar cycle. The average menstrual period in women is 28 days, close to the synodic period of the Moon, and this may not be a coincidence, pointing to an intimate connection between geophysical events and biology. Indeed, as I explain below, in the past the Moon was closer to the Earth, and theoretical studies and geologic evidence indicate that, 500 million years ago, at a time when biological clocks might first have been set in primitive living things, the synodic period of the Moon was 28 days. Nights are brightest and tides are highest, the so-called spring tides, every full Moon because of the reinforced tidal effects of the Moon and the Sun, since at this time both are aligned with the Earth. One wonders about the origin of this intriguing relationship which seems to point to a distant past when simple life forms living in tidal pools were directly affected by these geophysical events.

Although, in cosmic terms, our solar system is a tiny place, a mere speck of microscopic dust, it is quite large in our terms. If you traveled in an airplane for two days without stopping, you would be able to go once around the Earth's equator, while to get to the Moon, traveling at the same speed, would take just twenty days. Look at the Moon when it is visible during the day, an impressive sight because it just seems to hang in the sky quite detached, and think as you look at it that you could fly to it and reach it in twenty days. You will then get some idea of how far it really is. If you continued for another 20 years, you would reach the Sun. To get to Pluto, at the outer reaches of the planetary system, you would have to continue for about 800 years. If you kept on going, you would reach the nearest star system at Alpha Centauri, in approximately 1000 million years, about a quarter of the age of the Earth, and you still would not have gotten very far. Light, traveling at its fantastically high speed, can do this in just four years.

Thirty years ago, on July 20, 1969, at a place on the Moon called Tranquility Base, following a voyage that took three days to traverse the distance of 240000 miles (384000 km), NASA's Apollo 11 astronauts, Neil Armstrong and Edwin Aldrin, became the first humans to walk on the Moon. Astronaut Michael Collins, waited in the Lunar Module in orbit around the Moon, to bring them back safely to Earth. Behind the commander, the spirit of Isaac Newton guided the spacecraft. This was, beyond question, a historic moment for our species, *Homo sapiens*, who had surely come a long way, and I don't mean just from the Earth to the Moon. If you saw the great initial scenes of Stanley Kubrik's film *2001 A Space Odyssey*, you know what I mean. A total of seven missions manned by twenty-one astronauts went to the Moon, the last, Apollo 17, arriving there in December 1972. (Apollo 13 was not able to land because of an accident.) Twelve lucky human beings walked on the surface of the Moon, a magnificent adventure, indeed "a giant leap for mankind."

The Moon has only about 1/80 the mass of the Earth, and is also smaller, so that you are closer to its center when on its surface. Thus, your weight there would be different from your weight here. Should you weigh a hefty 240 pounds here, your weight there would be only about 40 pounds. This allowed the Apollo astronauts to make great leaps and bounds despite wearing heavy space suits. Although your weight on the Moon is less, your mass[1] – a measure of the quantity of matter in your body – is the same as on Earth. The Moon has been geologically dead for more than 2 billion years, because there is no atmosphere or water to erode its surface, it offers an indelible record of the history of this part of the solar system. This means it holds many secrets related to the formation of the terrestrial planets. Some of these secrets we have discovered, while others wait to be revealed some day to future explorers.

Thousands of photographs, about 800 pounds (360 kg) of Moon rock, and

---

[1] The weight in pounds is a measure of the gravitational force of attraction on the surface of the Earth, although in practice we use it as if it were a measure of the quantity of matter, as when we buy a pound of vegetables. Strictly speaking the quantity of matter is measured in a unit called "slug" in the British system (and kilogram in the metric system). When you buy a pound of beans what you really mean is 32 slugs of beans.

Apollo 17 reached the Moon on December 11, 1972. The Taurus-Littrow landing site is located between Mare Serenitatis and Mare Tranquillitatis (the landing site of Apollo 11). Astronaut–scientist Harrison Schmitt examines a large boulder, on this, the last manned mission to the Moon. (NASA)

the results of many experiments conducted on the lunar surface, were the treasures returned by the Apollo explorers, of greater value than all the gold and silver stolen from the Indians by the Spanish conquistadors. This material is the basis of our modern understanding of the formation and history of the Moon and the Earth. Perhaps the most important finding was that the overall chemical composition of Moon rocks was quite similar to that of Earth's surface, but with fewer volatile materials (those with low melting points) and more refractory elements (those with high melting points). It is as if they were rocks of Earth heated to high temperatures. The Moon's density is about 40 percent lower than Earth's, showing that it either lacks an iron core like that of the Earth or that its core is very small.

The exploration of the Moon also settled the question about the origin of the lunar craters. They are the results of impacts which over the ages have left their marks upon the lunar surface, and have remained there for us to study because the Moon is not geologically active. This is as expected and backs up the idea that the formation of the planets was the result of countless collisions between objects growing in the solar nebula. If you look at the full Moon with binoculars, you will clearly see Tycho, one of the most conspicuous craters on the lunar surface, on its southern hemisphere. It is 50 miles in diameter, 3 miles deep, and the central peak, typical of many impact craters, is more than 1 mile high. It is likely that Tycho is the most recently formed of the large craters on the Moon, with an age estimated to be only 100 million years. A system of rays runs from the crater in all directions, a mark left by the material ejected after the impact of an object of about 3 miles in size. To the north, clearly visible in the middle of one of the maria – the dark areas where few craters are found – is another large crater named in honor of Copernicus.

It is currently thought that the Moon formed as the result of a titanic collision between the Earth and an object the size of Mars at a time when the Earth was forming about 4.5 billion years ago. Perhaps we should pause here and reflect about this age which is almost beyond comprehension. We are talking about an event that happened 4500 million years ago. We can measure this number by the technique of radioactive dating applied to lunar rocks. We can write this number down and talk about it, but it is quite difficult to imagine because it is difficult to relate it to our experience, which is unfortunately very much shorter. With great luck your stay on Earth might be 100 years and just trying to imagine *one* million years is quite difficult, this being 10 000 times longer. If you could travel back into history in a time capsule (as they do in the movies) and you could travel one year every second, then to visit the Cretaceous period and say hello (from a distance) to *Tyrannosaurus rex* you would exit your capsule about six years later. To go all the way back to the birth of Earth, you would have to travel for 150 years. We are talking about very, *very* long time spans. Our species, *Homo sapiens*, has been around for roughly 200 000 years, quite a long time in terms of our life span, but in geological terms we have only just arrived. *If the history of the Earth to date was a long three-hour movie, then our species appears just in the last second.* If you blink your eye, you might miss us.

The grazing collision which produced the Moon generated so much energy that it vaporized a large fraction of the Earth's surface material and that of the impactor, in the process also giving the Earth a substantial spin-up so that the length of the day became only a few hours. The collision also tilted the Earth's axis of rotation so that it is not at a right angle to the plane of Earth's orbit around the Sun. This created the seasons, an important feature of our environment. Earth's surface material, melted by the enormous heat produced by the collision, was blasted into space and formed a

On December 7, 1992, during its journey to explore the Jupiter system, the Galileo
spacecraft took this image of the Moon. The prominent Tycho impact crater can be seen at
the bottom of the image with distinct bright rays extending radially outward. The dark
areas (marias) are lava-filled impact basins. Between Oceanus Procellarum (on the far left)
and Mare Imbrium (center left) you can see the prominent crater Copernicus. The crater to
the left of Copernicus is Kepler. The conspicuous round dark crater at the northern edge of
Mare Imbrium is Plato. Mare Serenitatis, Mare Tranquillitatis, and Mare Fecunditatis are
the three large circular areas which run from the center diagonally to the lower right (top
to bottom respectively), and Mare Crisium is near the right edge. (NASA/JPL/Caltech)

disk in orbit around the Earth. In the same way the planets formed from the
disk of material surrounding the Sun, the Moon coalesced from this disk.
This not only explains the results obtained from the lunar explorations, but
also the fact that the Moon is slowly moving away from Earth. The distance
to the Moon was measured very precisely after the Apollo 12 astronauts
placed a reflector on its surface. A laser beam from Earth was reflected back
from this providing the Earth–Moon distance, which was found to be
increasing by 1½ inches each year. This does not sound like much but it

About 4 billion years ago, a titanic collision occurred between an object the size of Mars and the young Earth as depicted in this illustration by Don Davis. The off-center collision generated so much energy that it vaporized a large fraction of the Earth's surface material and that of the impactor. The melted material was blasted into space and formed a disk in Earth orbit. In the same way that the planets formed from the disk of material surrounding the Sun, the Moon coalesced from the disk of material surrounding the Earth. (Illustration by Don Davis reproduced by permission of Sky Publishing Corporation)

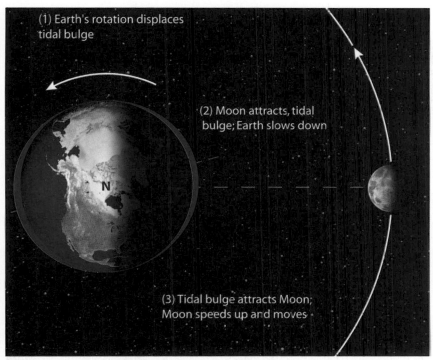

The ocean tides (highly exaggerated in this sketch) are dragged ahead of the Earth–Moon line (yellow) by the rotation of the Earth so that they are aligned along the red line. The bulge nearer the Moon feels a slightly larger force than the bulge farther from the Moon, the result being a net force (really a torque) that acts to slow down the rotation of the Earth. At the same time the Moon will feel the pull from this resultant force, gradually speeding up in its orbit and causing it to move away. If by chance the Moon moved in a direction opposite to Earth's rotation, then the forces would act in such a way as to speed up the rotation of the Earth, and bring the Moon closer, until an eventual and catastrophic collision. (José F. Salgado)

means that a few thousand million years ago the Moon was much closer to Earth than it is now.

As we have seen, the gravitational force of attraction between two massive bodies is larger for larger masses and increases as the distance between them decreases. Tides result from the difference in the Moon's gravitational force between two opposite sides of the Earth. The side nearer to the Moon is pulled with greater force than the center of the Earth while the side farther from it is pulled with less force. When the centrifugal force, which is equal and opposite to the gravitational force on the center, is subtracted, the result is two forces pulling away from the center. This mostly affects the oceans which are fluid and therefore easily deformed by this tidal effect resulting in *two* bulges (on average of about 1 meter in height)

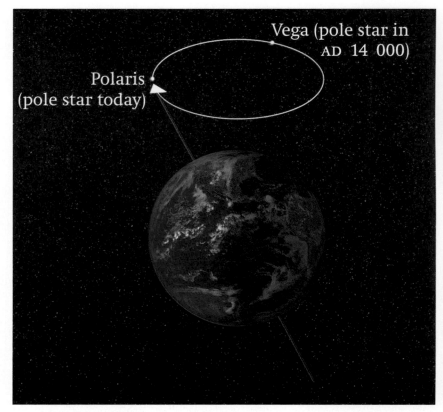

The axis of the Earth, like the one of a spinning top, will slowly wander in space, describing a circle every 26 000 years. It is not a smooth circle, since the axis also wobbles slightly during this motion therefore changing the inclination angle of the axis with respect to the ecliptic. Whereas it now points at the star Polaris (the reason for its name), in about 12 000 years it will be pointing at the star Vega. This motion of the axis is called precession, and it makes the position of the Sun, at the time of the equinoxes, slowly change with respect to the background stars. (José F. Salgado)

on opposite sides of the Earth in line with the Moon. As the Earth rotates under these bulges, two high tides will be experienced at any location in a day, although owing to the motion of the Moon, the time between them is closer to 13 hours. Because of the rotation of the Earth and the fact that the oceans cannot respond immediately to the changing position of the Moon, the tidal bulges are dragged *ahead* of the Earth–Moon line, and the force of the Moon on these bulges slowly acts to brake the rotation of the Earth. The equal and opposite forces exerted by the bulges on the Moon will speed it up in its orbit so that the Moon moves away, and the synodic period slowly increases.

The rapid rotation that existed right after the collision has now become a 24-hour day. It has been determined, by comparison with extremely accurate atomic clocks, that the length of our day increases by about one-thousandth of a second in 100 years. No need to adjust your watch, but over 1000 million years this can accumulate to a few hours.

Our Moon causes the spin axis of the Earth, which is inclined by 23.5 degrees to the plane of the Earth's orbit about the Sun, to slowly precess, that is, to move about in space describing a cone, completing one turn in about 26 000 years. If you have ever played with a top you have seen a pre-cession. Thus, whereas the Earth's rotation axis currently happens to point at the star Polaris, (the name is no coincidence) it will point at the bright star Vega in the constellation Lyra about 13 000 years from now. However, the inclination of 23.5 degrees will be maintained with only small changes. If it were not for the gravitational pull of the Moon the axis of the Earth would shift its direction in space over a large range and this would produce enormous climatic changes over time scales of tens of millions of years. This stabilizing effect of the Moon on Earth's rotation has been crucial to the development of life on Earth.

Fertilization of Earth

At the time the terrestrial planets formed, the nebula in this region was hot enough to prevent the condensation of large quantities of the volatile elements. The proto-Earth, hot and partially melted because of both its high internal heat of formation and the repeated impacts by planetesimals, was a sterile place. Yet today the Earth has a significant quantity of water, about $\frac{1}{4000}$ of its total mass. Although this might not sound much, if you took all the water on Earth it would fill a sphere as large as one half the Moon. The question to be answered then is where all this water came from, allowing the sterile Earth to become fertile, and eventually to spawn life. We also know, as we shall see in Chapter 5, that life arose quite soon after the initial bombardment ceased, raising the question of the origin of the organic materials used by life.

The answer is cast in stone on the pockmarked surface of the Moon, which after the Apollo expeditions we know to be the result of countless impacts by objects which early in the history of the solar system continued to bombard both the Moon and the Earth. The giant planets, Jupiter, Saturn, Uranus, and Neptune, formed after the sterile and dry terrestrial ones, in a region where icy planetesimals were abundant. The gravitational force exerted by the large planets on the small planetesimals often deflected them from their orbits, and flung them out into the Oort cloud or toward the inner parts of the solar system. In this way, millions of these small, icy planetesimals, a torrential rain of objects, often collided with the terrestrial planets, until the sky cleared. This is not just a pretty analogy, since these planetesimals, formed in the outer regions of the solar nebula, contained a large quantity of water, which contributed to the formation of oceans on Earth. It was this final storm, which lasted a few hundred million years and ended about 3.8 billion years ago, which brought to the Earth the missing ingredients for life, including complex organic molecules which were first formed on the surface of tiny interstellar grains. Of course, this bombardment also hit the Moon and must have hit all terrestrial planets. In fact the surface of Mercury looks very much like that of the Moon, with craters everywhere. However, the different physical conditions on each of these bodies, being of different masses and sizes, and at different distances from the Sun, meant a very different outcome.

The escape velocity from the surface of a planet is the velocity an object needs, be it a rocket or a gas molecule in the atmosphere, to overcome the gravitational force exerted by the planet and escape into deep space. It is also the arrival velocity of any object, like a meteor, if it started from rest, very far away attracted by the gravitational pull of the Earth. We will use this fact in Chapter 7. For the Earth, this velocity is about 7 miles per second or about 25 000 miles per hour, close to the escape velocity for Venus since these two planets are of similar mass and size. Objects which do not achieve this velocity will fall back to Earth although they might reach great heights before doing so. The much lower mass of the Moon means a lower escape velocity of only 5000 miles per hour. On Mars and Mercury, larger than the Moon but smaller than the Earth, the escape velocity is about twice that of the Moon.

In a gas such as the Earth's atmosphere, the atoms and molecules are constantly moving in all directions with many different velocities. The average velocity of the molecules or atoms is determined by the temperature, being higher for higher temperatures. At a given temperature, the lighter molecules move faster, so hydrogen moves faster on average than its heavier isotope deuterium, and much faster than oxygen. Thus hydrogen can more easily reach escape velocity, and escape a planet. Of course the gas molecules in an atmosphere will have a wide range of velocities, some much higher than the average, and these will be the ones to escape. On a hot day, the average velocity of an oxygen molecule in air is about 2 miles per second while for a lighter hydrogen molecule it is 8 miles per second. Consequently it is easy for hydrogen molecules in our atmosphere to escape, and its rate of depletion can be measured. In contrast, oxygen does not escape as easily.

What really matters is the temperature at the top layers of the atmosphere, since that is the region from which gas can escape a planet. This so-called exosphere begins for the Earth at a height of about 300 miles.

Because of its small mass, gravity on the Moon is only one-sixth of that on Earth, too weak a force to trap the volatile elements vaporized in a collision, so they escaped into space instead of forming oceans and atmospheres. Although the escape velocity on Mercury is twice that on the Moon, the temperature is so much higher there that Mercury also was never able to retain an atmosphere. In contrast Mars, although of small mass, is sufficiently far from the Sun that it is very cold and still possesses a tenuous atmosphere, the remnant of what initially was a much thicker one. Venus, being closer to the Sun than the Earth, receives about double the solar energy. Its surface temperature, even at a time when the young Sun was less bright than today, was so high that water never condensed out of its atmosphere. This atmospheric steam was broken up by solar ultraviolet radiation into its component atoms, hydrogen and oxygen. Most of the light hydrogen easily escaped into space. The oxygen combined with other atoms to form various oxides, and combined with sulfur and some remaining hydrogen to form the sulfuric acid that we find today in the clouds of the Venusian atmosphere. On Earth, geochemical processes removed atmospheric carbon dioxide but these processes depend on liquid water, and so never operated on Venus. So, although Earth and Venus probably started out very similar to one another, they developed very differently. Today the surface of Venus might best be described as Hell.

Liquid water, the most important compound for the development of life, can only exist if it is under some pressure like that provided by an atmosphere. With less pressure it will boil at lower temperatures, something that even affects the time needed to cook a hard-boiled egg on top of a mountain. On the surface of a planet without an atmosphere, no oceans can exist, although solid water – ice – can, if the temperature is low enough. In fact, it was a surprise to find evidence for ice inside some craters on the poles of Mercury, which because of their location are in permanent shadow and are therefore extremely cold areas despite the nearness of that planet to the Sun. On Mars it is quite likely that the bombardment of planetesimals contributed enough volatile elements to create oceans and an atmosphere of carbon dioxide. Today we see the evidence for this in the form of ancient river beds traversing the surface of Mars, and an absence of craters in some regions of the northern hemisphere, which might have been under water long ago. The cold poles of Mars are covered with ice, composed of frozen carbon dioxide and the remnant of ancient waters. Early observers of the

On March 11, 1997, Alessandro Dimai, a member of the Associazione Astronomica Cortina, took this photograph of Comet Hale-Bopp near Cortina d'Ampezzo, in the Dolomite mountains of northern Italy. The bluish gas tail (the color due to the emission by ionized carbon monoxide) is pushed away from the comet by the solar wind, and the brighter dust tail (its yellow color from reflected sunlight) is driven by the slight pressure of sunlight. Both tails point away from the Sun – they do not trail the motion of the comet. (Alessandro Dimai)

Moon believed that the dark crater-less areas there were indeed oceans of water and called them maria. In some sense the maria *are* oceans, the solidified remains of oceans of lava which long ago covered a good fraction of the lunar surface.

Even today, the rain of planetesimals which began 5 billion years ago has not completely stopped. It has turned into a very light drizzle of planetesimals which sometimes approach the inner solar system from beyond the orbit of Pluto. They are recognized by the evaporating gas and expelled dust which become visible as they encounter the solar wind and feel the solar heat. A fuzzy halo with a long magnificent luminous tail develops: a comet. Although comets were the ultimate source of the material for life, other conditions were important for its establishment and development. It was of little use if the ingredients for life were delivered and then promptly dissipated into space because of the low gravitation of the object, as on the Moon, or because of the very high temperatures on the surface, as on Mercury and Venus. We shall see later that comets can on occasion even become the ultimate source of death.

Perhaps, over many millions of years after their formation, life could have emerged on Mars and Venus. However, the development of their environments led to conditions which today do not seem appropriate for the support of life as we know it. Only on Earth, located in the so-called habitable zone of our Sun, were the conditions favorable for the emergence of life and, more important, over the several-billion-year history of Earth its environment changed in such a way that life could adapt to these changes and prosper. Life also contributed to these changes, to which it then had to adapt. We are concerned today with the drastic impact that human activities are having on the global environment, causing changes which might be so rapid that life may not be able, or have the time, to adapt.

Let's take a more detailed look at our planet in the next chapter.

In the habitable zone of the Sun, 93 million miles away from it, a beautiful planet, a blue and white jewel, sometimes with a greenish-yellow tint, and a faint atmospheric halo, is found: *Earth*. It is the largest of the four terrestrial planets and the only one with a large moon. In this image, produced with data obtained by surrogate eyes on several satellites, we see the side of Earth that is illuminated by the light of the Sun, which provides the energy for life. Three-quarters of the surface is covered by water. The storm raging off the west coast of North America is hurricane Linda. You can see the mighty Amazon river crossing the last great rain forest and transporting thousands of tons of sediment into the Atlantic, and the crumpled west edge of South America rising to form the Andes. A thin atmospheric halo surrounds the globe. The Moon, at double its apparent size, is an artistic addition. (NASA GSFC Scientific Visualization Studio)

Chapter 4

Mother Earth

Orbiting this at a distance of roughly ninety-eight million miles is an utterly insignificant little blue-green planet . . .[1]

Our home is a unique planet in a very special place of the solar system, a place where liquid water can be present on its surface. It is a dynamic planet, having gone through great transformations during its 4.5-billion-year history. Wherever we look, we find that life and the environment are inter-related in a complex and balanced way. An intricate web of geophysical and biological processes maintains this equilibrium, which is fairly stable, but not necessarily permanent.

Earth

The "habitable zone" around a star is that region surrounding it where the temperature on the surface of an orbiting planet can be above freezing and below the boiling point of water. This means that such a planet can have liquid water, something we understand is essential for life as we know it. In the habitable zone of the Sun, 93 million miles away from it, a beautiful planet is found: *Earth*. The habitable zone is not very wide: move Earth toward the orbit of Venus and it becomes hell; move it nearer to the orbit of Mars and hell freezes over.

Over our lifetimes, indeed over all of human history, the Earth seems to be a reassuringly stable place giving us some solid ground for comfort (pun intended). Yes, there are the occasional earthquakes, tsunamis, volcanic eruptions, hurricanes and tornadoes, but on the whole Earth seems fairly changeless. However, this is only an illusion caused by our extremely short stay on this planet. The Earth has changed significantly on a time scale of millions of years: continents have changed position, deep ocean basins have opened where continents separated, tall mountain ranges and volcanos have surged where continents collided, and the climate has changed.

[1] Douglas Adams, *The Hitchhiker's Guide to the Galaxy*, Six Stories, p. 5.

Taken as a whole, the Earth is a sphere with a diameter of about 8000 miles (13 000 km), mostly composed of just a few elements: 35 percent iron, 30 percent oxygen, 15 percent silicon, and 13 percent magnesium, leaving only 7 percent for all the other elements which make up this complex and wonderful world. Most of these elements are bound into the great variety of minerals which are found on, and inside, the Earth.

That our Earth is a sphere was already deduced by those who thought about these things in antiquity, but the voyage of Magellan (see page 28) provided the proof. If that were not sufficient, then photos from space should do. Some people prefer to believe that the Earth is flat, that we have never been to the Moon, and that the world was created in 4004 BC and will come to an end any day now. But the truth, although perhaps more intricate, is so much more interesting that I cannot understand what is to be gained by ignoring it.

The orbit of the Earth around the Sun is not quite a perfect circle but is slightly eccentric or egg-shaped – an ellipse, as first determined by Kepler – so that over a year the distance to the Sun varies from about 91 to 94 million miles (147 to 152 million km). In our model from the first chapter, the distance between the Earth and the Sun would go from 690 feet to 710 feet, not a great change as you can see. Many wrongly believe that this is the explanation for the seasons thinking that when we are nearer the Sun it is summer and when we are farther away it is winter. However, as you probably know, when it is summer in the northern hemisphere it is winter south of the equator, so the distance to the Sun cannot possibly explain the seasons. At any rate, the distance only changes by 3 percent, not enough for the observed variation in the climate. In fact, the seasons are a result of the Earth's axis being tilted from the plane of its orbit around the Sun (by 23.5 degrees). As illustrated in the figure, this changes how much solar energy falls on each hemisphere of the Earth over the course of a year, raising or lowering the average temperature accordingly. Because of this tilt the highest point reached by the Sun over the horizon at a particular location changes over the course of a year. Twice a year, on March 21 and September 22 – the equinoxes – the Sun is positioned directly over the equator and daytime and nighttime are of equal length everywhere. How different Earth's climate would be if the Earth's axis were not inclined. Because the days and nights would always be of equal length (12 hours) and, at a particular point on Earth, the Sun would always reach the same height no matter what time of year – lower as you moved away from the equator – there would be no seasons. The entire planet would be slightly warmer over the part of the year when it is closer to the Sun, because of the eccentricity of its orbit.

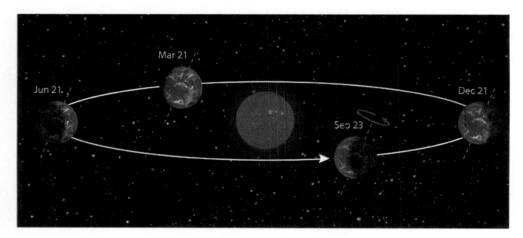

Two factors influence the average amount of sunlight received by the ground and therefore its temperature: the length of daylight (longer daylight means more heating) and the angle at which sunlight strikes the surface. When sunlight strikes a surface at right angles it is more concentrated over the surface than if it strikes at an angle. So at noon and in the summer, when the Sun is higher from the horizon, there is more heating. The seasons are caused by the changing average amount of sunlight the ground receives due to the tilt of the Earth's axis with respect to the plane of its orbit as it travels about the Sun. This tilt is maintained with respect to the stars, that is the axis always points in the same direction (towards Polaris) as illustrated, and so, on December 21 the north end of the axis points away from the Sun and it is winter in the north and the north pole is in darkness. Six months later, on June 21 the situation is reversed. When it is summer in the northern hemisphere it is winter in the southern one. On September 23 and March 21 (the equinoxes), day and night are of equal length. (José F. Salgado)

There would be no spring to celebrate the rebirth of life and no autumn to become melancholic. As we saw in the previous chapter, the Moon stabilizes this tilt and so has stabilized Earth's climate over its history, an important ingredient for the evolution of life. It is believed by some to be so important that it might be fitting to describe ourselves as Lunatics, a name which I think is right for other reasons.

Although stable, the Earth wobbles slightly in space so that the tilt of its axis slowly changes between 22 and 25 degrees in a cycle of about 40 000 years. Significant climatic changes in the past are related to this change in the tilt of the Earth's axis, since a greater tilt will result in greater seasonal variations. The last great ice age, only 15 000 years ago, covered one-third of the land with an ice sheet a mile thick. It is astonishing that it took a change of only a few degrees in the average temperature for this to happen. We are used to small changes having small consequences: press the accelerator slightly and the speed of your car will increase slightly, increase the fire in your oven slightly and the temperature will increase slightly. However, for complex systems like our Earth this is not necessarily so: a small change in one process of an intricate chain can affect several subsequent steps in the chain so that they all add up and result in a large final change.

Compared with other planets, ours has the greatest variety of climates and environments, because it is in the habitable zone, which allows for this. Also, it is geologically active and, over the long time span of its history, high mountain chains and deep oceans have emerged and later vanished. Mount Everest in the Himalayas is the highest mountain on Earth, 29029 feet (8848 m) high. The tallest mountain in Africa is Kilimanjaro (19340 feet, 5895 m) with its snow-covered summit only 3° south of the equator. The oceans are deepest at the Marianas Trench in the Pacific, the depth there being 36198 feet (11033 m), followed by the deepest part of the Atlantic in the Puerto Rico Trench at 30246 feet (9219 m). For us, who measure only about 6 feet, these are incredibly high and deep places but, from a different perspective, if the Earth were the size of a basketball, its surface would be smoother than that of a real-life basketball. Temperatures as low as −89 °C (−129 °F), have been recorded at the Russian Vostok research station in Antarctica, near the south pole, while a record high of 57 °C (134 °F) has been registered in the aptly named Death Valley of California, the lowest point in North America.

Mount Everest, the highest mountain on our planet, 29029 feet (8848 m) high, is known to the Tibetans as "Goddess, Mother of Earth." It is located in the Himalayas on the border between Tibet and Nepal. The challenge to reach its summit was met on May 29, 1953, by Sir Edmund Hillary (1919–) of New Zealand and Tenzing Norkay (about 1914–1986), a Nepalese mountain climber of the Sherpa tribe. This photograph was taken from the Space Shuttle on a morning, and shadows are west of the high features. Mount Everest, at the center of the photograph, casts the largest shadow. Several large, impressive glaciers radiating from the mountaintop occupy the rugged valleys. (NASA)

Earth is not a homogeneous body with the same composition through-out; it is a "differentiated" planet, composed of concentric shells with different properties. After the heavy initial bombardment from space during the formation of our planetary system, the Earth was a very hot place, a molten mass of material. Just as in a blast furnace molten metal sinks to the bottom and slag floats on top because of their different densities, the material of Earth separated into layers, with the heaviest metals at the core. Without this differentiation, and without the process of plate tectonics which led to the formation of continents, oceans, and the atmosphere, life would probably have developed differently, if at all. Going from the surface to the center of the Earth, both the temperature and pressure increase. The central, hot, iron–nickel core is a gigantic metal sphere with a diameter of about 4000 miles (6500 km). This is about the size of Mars. It contains about one-third of Earth's mass and is very different from the material surrounding it. The temperature at the center of the Earth is still quite high, about 6000 °C, coincidentally as hot as the surface of the Sun. Although the melting temperature of iron is much lower (about 1500 °C or 2700 °F), the inner half of the core is solid metal because of the high central pressure exerted by the outer layers. The outer half of the core, where the pressure is lower, contains molten metal. As the Earth spins, this molten metal generates a magnetic field, converting Earth into a gigantic bar magnet used by birds and navigators to orient themselves.

The reason the Earth's core remains hot after all these years is partly due to remnant heat of the formation process, but mostly due to the decay of some radioactive isotopes with long half-lives, such as uranium, thorium, and potassium, which were incorporated into the Earth at the time of its formation. Our Earth would be a very different place, geologically dead, if it were not for the heat provided by these radioactive elements which formed in the very distant past when a massive star exploded as a supernova. The Earth's mantle, which is about 1700 miles (2800 km) thick, surrounds the core. It is composed of semi-solid rocks of intermediate density containing compounds of oxygen with magnesium, iron, and silicon. The Earth's thin outer layer, about 60 miles (100 km) deep, the lithosphere (from the Greek lithos, "rock"), is composed of light materials containing primarily compounds of oxygen with silicon, aluminum, calcium, magnesium, sodium, and potassium. It is divided into several large slabs or plates of solid rock which cover the globe. The outermost region of the lithosphere carries the continental crust which is 15 to 30 miles (25 to 50 km) thick, and the thinner and denser oceanic crust, 3 to 6 miles (5 to 10 km) thick, which makes up the bottom of the oceans. Under the rigid lithospheric plates lies a layer called the asthenosphere (from the Greek asthenic, "weak") which is partially molten and therefore not rigid.

The structure of the Earth. (José Salgado, after a diagram by USGS)

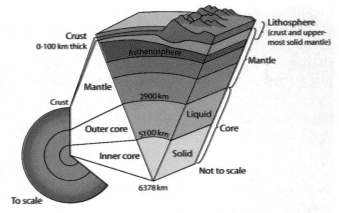

Air

Lying on your back on the pristine white sand of that beach of Chapter 1, you will enjoy gazing at a beautiful light-blue sky and maybe some puffy whitish clouds of water vapor, which, as if they were some kind of ethereal sheep, slowly drift in the air. You are looking at the atmosphere, a thin layer of gas surrounding our planet, which contains the oxygen we need to breathe. It also keeps the planet's surface warm and shields us from harmful ultraviolet solar radiation and cosmic rays, and from being bombarded by small cosmic projectiles (the meteors) which enter it at high speeds. Unfortunately, we also use our atmosphere as a garbage dump.

The atmosphere plays an important role in the regulation of energy which arrives from the Sun to heat the Earth's surface, and the flow of energy from the surface into space, which cools it. The reflectivity of the surface (the albedo) determines how much energy is returned to space: the more energy it reflects back into space, the less it will absorb. It is the balance between these opposite flows which determines the average temperature on the surface. When, for example, the tilt of the Earth's axis increases, winters become colder, so there are larger snow-and-ice-covered areas for a longer fraction of the year. These will reflect a larger fraction of the incoming sunlight back into space, instead of absorbing it, and so the delicate balance which regulates the climate will be disturbed. This will lead to an overall cooling of the surface and could produce an ice age. You use this same property when, in the summer, you keep cool by using light clothing.

When you feel the heat from the Sun, it is because your skin is absorbing part of the energy transported by sunlight, which is a form of electromag-

A thin layer of gas composed mostly of nitrogen and oxygen surrounds our planet. It provides the air we breathe, keeps our planet's surface warm, protects us from damaging ultraviolet radiation, and shields us from small projectiles. Unfortunately we also use this delicate layer as a garbage dump. In this image of the rising Sun, as seen from the STS-101 mission of the Space Shuttle, sunlight is refracted into bands of different colors by the atmosphere. You can appreciate the thinness of our atmosphere by comparing it with the slight curvature of the Earth's horizon and mentally completing the globe of the Earth. (NASA)

netic radiation. This travels at the speed of light which, as we have seen, takes only 8 minutes to arrive on Earth. What distinguishes different kinds of electromagnetic radiation is the wavelength, or equivalently the frequency, of the vibrating electric and magnetic fields which make up this electromagnetic wave. This was first worked out by another great hero of the natural sciences, the Scottish physicist and mathematician, James Clerk Maxwell (1831–1879), who joins Newton and Einstein as one of the greatest physicists of modern times. His mathematical relationships – Maxwell's equations – are the starting point of all calculations dealing with modern electronic devices. Electromagnetic waves bring us news from the most distant regions of our Universe, received by large optical and radio telescopes on Earth, or X-ray and ultraviolet telescopes in space. The same type of waves also bring news from your local radio or TV station. So, next time you see your favorite sports event live on TV, thank Maxwell for his equations.

The wavelength of visible light is very small, about 500 nanometers (nm), where a nanometer is a convenient unit equal to a billionth of a meter. Red light, for example, has a wavelength of about 650 nm. Blue light has a shorter wavelength, around 475 nm, and invisible ultraviolet radiation has an even shorter wavelength. Infrared, on the other hand, has a wavelength longer than red light. The use of the prefixes "infra" and "ultra" refer to the frequency of light, which is the inverse of the wavelength (the longer the wavelength, the lower the frequency), so that *infra*red light has a *lower* frequency than red light.

Sunlight is most intense at visible wavelengths between 400 and 800 nm, although a fraction of the Sun's energy is produced at other invisible wavelengths such as ultraviolet and infrared. As it passes through our atmosphere, the light from the Sun is partly absorbed and scattered by different component gases, so that the composition of the light reaching the Earth's surface is different from that produced by the Sun, with important consequences for biology. This is why the color of the Sun as seen from the Earth is different from that seen from space: it has more blue light and so it looks yellower.

The term "visible" does not imply a special property of the electromagnetic wave arriving from the Sun, but rather a particular property of our light-sensing organ, the marvelous eye. It is sensitive to light with wave-

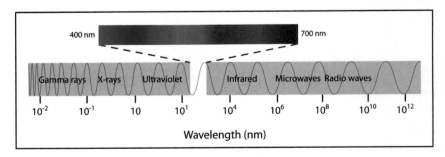

The electromagnetic spectrum. Visible light is only a tiny portion of the electromagnetic spectrum of waves which go from short-wavelength gamma rays to long-wavelength radio waves. One nanometer (1 nm) is one-billionth of a meter. (José F Salgado)

lengths between about 400 nm (violet) and 700 nm (red), and most sensitive to yellow-green light, a useful property if your main interest is finding and eating green things. It is not surprising that our eyes are sensitive to the most intense wavelengths of light arriving at the surface; they would not be very useful otherwise. Presumably on a planet around a star different from the Sun, say a smaller, cooler one, which shines mainly in the infrared region, eyes would be sensitive to infrared light.

An efficient way to obtain energy is to absorb it from an electromagnetic wave. This is what happens in a microwave oven, a device that generates electromagnetic waves at a wavelength which is easily absorbed by water molecules. Since water makes up most of the composition of anything you might want to cook and eat, this works well. Photosynthesis in plants, the principal source of energy for living things, functions by absorbing mostly red and blue light. Most of the things we see, we see by reflected sunlight shining from them. If something absorbs red and blue colors, removing them from sunlight, what remains will look green, and this is the reason vegetation has this color.

When sunlight strikes molecules or small dust particles in the atmosphere these scatter it, that is, deflect it in all directions. Scattering depends on the wavelength of light in such a way that blue light is scattered much more than red light. This is why the sky is blue. If instead we look at the Sun directly, more of the red light will reach us, since the blue light has been scattered away, and the Sun looks redder than it really is. The more gas there is in the path of the light, such as when the Sun is low on the horizon, the redder the Sun will appear. If there is some dust or smog in the atmosphere the effect will be still more dramatic, giving us the spectacle of a gorgeous red sunset. Not so gorgeous is the brown haze which you can sometimes see as you look at the horizon, again because your line of sight crosses a lot of air. If you look up instead the air seems clear, but it is just as polluted as the air that you see deceivingly far away towards the horizon.

On the Moon, where there is no atmosphere, the sky is black, with all stars visible, just as at night on Earth, except that they do not twinkle and, most surprisingly, they are seen even when the Sun is present. The Sun is a lighter shade of yellow than we see it. You saw the stars twinkle on the night you went out to gaze at them (did you?). This is because, as their light traveled through our atmosphere, it was bent first one way and then the other, as bubbles of turbulent air crossed the light's path.

The atmosphere thins rapidly as we move away from the surface: 99 percent of it is within about the first 20 miles above the surface and half of it is within 3 miles of the surface, lower than the highest mountains. It is composed of 78 percent molecular nitrogen (two atoms of nitrogen, N_2) and 21 percent molecular oxygen (O_2), with traces of carbon dioxide, water, and argon (Ar). A triatomic form of the oxygen molecule called ozone (O_3) is found in a thin layer in the lower stratosphere, between 9 and 18 miles (15 and 30 km) altitude. Four billion years ago, the atmosphere was very different, composed mostly of carbon dioxide, nitrogen, water vapor and sulfur dioxide (SO_2). This was also a time when our young Sun was about one-third less luminous than today. From this initial composition, our atmosphere evolved, gradually increasing its oxygen content, and decreasing its carbon dioxide content, because of photosynthetic processes in plants, and the removal of carbon dioxide with the build-up of carbon-based life, and carbon-bearing minerals.

Carbon dioxide, which is only a minute 0.0025 percent of the atmosphere, water vapor, and other so-called *greenhouse* gases act in the atmosphere to give the same result as the glass windows in a greenhouse, which is of course why they are called greenhouse gases. Although these gases are transparent to visible light, they are not to infrared radiation. Sunlight can therefore reach the surface of our planet where its energy will be absorbed, heating the surface. The heated surface will cool by radiating in the infrared wavelength region. Since the atmosphere is opaque to infrared wavelengths, it will trap this radiation and hence warm the lower parts of the atmosphere. The more greenhouse gases there are in an atmosphere, the warmer it becomes. The initial high carbon dioxide content in the Earth's atmosphere, 4 billion years ago, helped keep our planet warm and prevented it from freezing, compensating for the much lower luminosity of our young Sun.

Even today, without a warming atmosphere the surface of our planet would be lowered by nearly 30 °C (86 °F), becoming a cold, mostly frozen and inhospitable place. So life on Earth depends on the good services of a small quantity of greenhouse gases. However, as is often said, too much of a good thing can kill you. An increase in carbon dioxide in the atmosphere will cause the surface to become warmer, and a large increase could cause it to heat up excessively. Surface rocks, which contain carbon dioxide, could then release more of it into the atmosphere, increasing the temperature further, this again leading to more carbon dioxide and so on, a runaway situation. This is not mere theory, as we can see by looking at our "sister" planet Venus, which through this process has reached a surface temperature high enough to melt lead. This is why many scientists are working to understand the environmental changes caused by human activities, which include an increase in the generation of greenhouse gases, in particular carbon dioxide from the burning of fossil fuels. Large-scale deforestation aggravates this, by reducing the amount of carbon dioxide being removed from the atmosphere by photosynthesis. Studies of air trapped in Antarctic ice cores have shown that over the past 200 years the atmospheric carbon dioxide content has increased steadily. Other greenhouse gases such as methane (CH_4), produced by bacterial fermentation and by "bovine flatulence," a euphemism for farting cows, and nitrous oxide (N_2O), a byproduct of chemical fertilizer use and automobile exhaust, also contribute to global warming. Their long-term effects are, however, considered less important than those of carbon dioxide, because these gases do not remain in the atmosphere for as long.

Wherever we look, and the more we understand, we find that life and the environment are interrelated in a complex and balanced way. An intricate web of geophysical and biological processes maintains this equilibrium, which is fairly stable, but not necessarily permanent. A change of a small percentage, one way or another, in the rate of some of these processes could affect the environment and life in ways we cannot foresee because the chain of events is complex, and subtle effects are difficult to predict.

For example, as discussed further in Chapter 8, this is what happened to stratospheric ozone, where we suddenly realized that we were in deep trouble. A "hole" in the ozone layer over the poles, and a reduction of the global ozone concentration were discovered to be the result of the use of what were thought to be quite harmless chemicals.

In the lower stratosphere a series of reactions driven by solar ultraviolet radiation convert normal molecular oxygen into ozone, forming a delicate, and vulnerable, thin layer. If all the ozone in this layer were placed on the surface of the Earth, it would be only ⅛ inch thick. Ozone will absorb ultraviolet radiation and be separated into molecular and atomic oxygen, through a cycle regulated by a fragile balance. These reactions effectively absorb ultraviolet radiation, so that it does not reach the surface, providing a shield that protects life from this damaging radiation. It is curious that this gas, which protects life when in the stratosphere, is noxious on Earth's surface where a photochemical reaction in smog produces it. Its great reactivity damages plant and animal tissue. The ozone layer, which has protected life on the Earth's surface for more than 500 million years, is a byproduct of the increase in atmospheric oxygen provided by photosynthesis. We must be careful not to damage this unique shield. Before the formation of the ozone layer, living creatures had to live under water, where ultraviolet radiation does not penetrate.

We need oxygen to breathe, and it also helps reduce damage from incoming meteors, because as they enter the atmosphere they heat up owing to the great friction with air, and then burn at a certain temperature. However, too

"HONEY, I THINK YOU BETTER PUT ON SOME SUN-SCREEN."

much oxygen could also become a problem. Every year, many million lightning flashes hit the ground all around the world causing forest fires, creating power failures, and killing hundreds of people. It has been estimated that a small increase in the oxygen content of the atmosphere from 21 to 25 percent would cause vegetation to catch fire easily after a lightning strike, and become a major problem.

Every year, lightning flashes hit the ground all around the world, causing forest fires, creating power failures, and killing hundreds of people. They are, however, spectacular to watch. The speed of sound in air is 340 m/s (1100 feet/s), almost 1 million times slower than the speed of light. This is why, although lightning generates thunder, you see the light before you hear the thunder. This photo shows a beautiful lightning flash over Cabo Rojo in Puerto Rico. The planet Jupiter is the white dot at the right edge of the photo. (Frankie Lucena, Cabo Rojo, Puerto Rico)

History

From a study of the biblical record, Bishop James Ussher (1581–1656), an Anglo-Irish prelate, calculated that the Earth had been created in the year 4004 BC. This view prevailed for a long time. Somehow he was even able to figure out that this had happened on October 23 of that year! In the nineteenth century, the age of the Earth became a topic of great debate between those who followed Ussher's biblical chronology and those geologists who understood that the observed sedimentation of rocks, the growth of mountains, and the gouging of deep canyons were processes which required immense times, estimated then to be between 100 and 400 million years. Those who followed Darwin's thesis, that evolution also needed a very long time, joined the geologists in their view. This view became known as "uniformitarianism," postulating that Earth had a long history of slow, gradual change, something expounded in *Principles of Geology*, published in 1830 and written by Darwin's friend Sir Charles Lyell (1797–1875). This became a most influential book.

Pioneering studies of the fossil record by the founder of the science of paleontology (science of ancient life), the French naturalist Georges Leopold Cuvier (1769–1832), a contemporary of Lyell, were published in 1825 in *Discourse sur les Revolutions de la Surface du Globe* (*Discourse on the Revolutions of the Surface of the Globe*). Cuvier used "revolutions" to mean catastrophic change. His studies led to the realization that the fossil record was discontinuous, showing species which had become extinct to be replaced by others. Cuvier concluded that the geologic and fossil records bore evidence for great past cataclysms not in agreement with the uniformitarian idea of slow, gradual change.

Those who stuck to the biblical chronology had no choice but to argue that the abrupt changes seen in the fossil record, and things like mountains and canyons, must have been the results of a few humongous recent catastrophes, the last one being associated with Noah's universal deluge. We know today that the Earth is tens of times older than the uniformitarians thought, there being ample time for slow geological processes such as erosion and sedimentation to modify the landscape. But we have also learned that there were catastrophes in the past and, most surprisingly, that we are the result of one such event, as we shall see in Chapter 6.

The Copernican revolution displaced humans from the center of the Universe, and as the immensity of the Universe revealed itself to astronomers, we became an increasingly insignificant part of it. Whereas a history of 6000 years or so was something we could relate to, billions of years of history were not, and made our lifetimes only an infinitesimal part of the history of Earth. Another horizon was being expanded, pointing out that humans were not only insignificant in space but also in time. The essence of the intellectual struggle of those days was our relation to the Universe. However, until a reliable measurement of the ages of rocks could be obtained, much of this debate lacked a solid scientific basis.

In the latter part of the nineteenth century, several attempts were made to compute the age of the Earth and Sun using a scientific approach based on physical laws, and not relying on circumstantial geological evidence. Foremost among the scientists working on this problem was the British physicist William Thomson (1824–1907), knighted by Queen Victoria to become Lord Kelvin in 1892, a most influential figure in the natural sciences of his time. Based on the assumption that the heat from the Earth's interior was a remnant from a time when it was initially formed, and the theory of heat to which he contributed significantly, Kelvin computed an age for the Earth of about 100 million years, quite a bit older than Ussher's but not old enough to fit the uniformitarian view. Kelvin also concluded that the Sun was not much older than 100 million years, that it had been much hotter 1 million years ago, and would continue to cool in the future, not providing sufficient time for the uniformitarian view of the evolution of life. Kelvin wrote in 1862:

It seems, therefore, on the whole most probable that the Sun has not illuminated the Earth for 100,000,000 years, and almost certain that he has not done so for 500,000,000 years. As for the future we may say, with equal certainty, that inhabitants of the Earth cannot continue to enjoy the light and heat essential to their life, for many million years longer, unless sources now unknown to us are prepared in the great storehouse of creation.[2]

And this is indeed what happened: there were sources unknown "in the great storehouse of creation." Not only did the discovery of radioactivity provide the missing source of heat allowing the Earth to be much older than computed by Kelvin, it also provided a precise clock to determine the age of rocks. Perhaps Kelvin's greatest contribution to the question of the age of the Earth was his insistence that this had to be established using the known laws of physics. The discovery of radioactivity and the processes of nuclear physics also led to an understanding of a star's energy source, and of the origin of the elements.

It was the celebrated physicist, Ernest Rutherford (1871–1937), born in New Zealand, and later a young professor at McGill University in Toronto, who determined that radioactive substances could provide large quantities of heat, a fact which invalidated Kelvin's conclusions. He presented these results in an invited lecture at the Royal Institution in London in 1904, and provides this account of his lecture:

> I came into the room, which was half dark, and spotted Lord Kelvin in the audience and realized that I was in for trouble at the last part of the speech dealing with the age of the Earth, where my views conflicted with his. To my relief Kelvin fell fast asleep, but as I came to the important point, I saw the old bird sit up, open an eye and cock a baleful glance at me! Then a sudden inspiration came, and I said Lord Kelvin had limited the age of the Earth, *provided no new source of heat was discovered.* That prophetic utterance refers to what we are now considering tonight, radium! Behold! The old boy beamed at me.[3]

As we have seen, the atoms ejected in a supernova explosion collide to produce a large variety of heavy elements, some of them radioactive. The atoms which do not collide fly into interstellar space and millions of years later strike the upper layers of our atmosphere. These "cosmic rays," mostly protons, the nuclei of hydrogen atoms, were first discovered by Victor Franz

[2] Quoted in Joe D. Burchfield, *Lord Kelvin and the Age of the Earth*, University of Chicago Press, 1990, p. 31.
[3] *Ibid.*, p. 164.

Hess (1883–1964), an Austrian physicist who in 1912 investigated the nature of the radiation. Using balloons filled with explosive hydrogen gas, he flew to a height of over 17 000 feet where, fighting the bitter cold and gasping for air in the thin atmosphere, he used instruments to investigate radiation in the upper atmosphere. Hess was awarded the Nobel Prize in Physics in 1936 for his discovery of cosmic rays. When these rays crash into the upper atmosphere, they cause reactions which transmute atoms of nitrogen-14 (^{14}N) into carbon-14, (^{14}C), a radioactive isotope of carbon. Since the carbon in living organisms is ultimately obtained from atmospheric carbon dioxide, the ratio of carbon-14 to normal carbon-12 in their tissues is the same as that in the atmosphere, about 1 in 10 000.

When organisms die, they no longer incorporate carbon into their tissues and, from this moment on, radioactive carbon-14 decays and becomes less in relation to carbon-12, which remains constant. Knowing that the half-life of carbon-14 is 5730 years, it then becomes possible to estimate the time that has passed since the death of the organism. If for example a certain fossil was 5730 years old, it would contain one-half as much carbon-14 in relation to carbon-12 as living tissues; if it were 11 460 years old (two half-lives) it would contain only one-quarter of the carbon-14. This is the basis for the radioactive carbon-dating method, used to measure the ages of such things as the famous Shroud of Turin, or the 5300-year-old Iceman found in the Otzi region of the Alps, on the Austrian-Italian border. Before the discovery of radioactivity in 1896 by the French physicist Antoine Henri Becquerel (1852–1908) it would have been impossible to determine these or any other ages precisely. Nor would it have been possible truly to know the age of the Earth.

The determination of ages from radioactive decay turned into a difficult and active area of research early in the twentieth century, and it soon became apparent that the Earth was more than 1 billion years old, something unthinkable until then. This was based on the study of radioactive elements with long half-lives, elements created in supernova explosions and incorporated into our planet as it accreted from interstellar grains. For example, uranium (^{238}U) decays to lead (^{206}Pb) with a half-life of 4.5 billion years, and potassium (^{40}K) to argon (^{40}Ar) with a half-life of 1.3 billion years. The oldest known rocks found in Greenland have an age of 3.8 billion years. Meteorites as old as 4.6 billion years have been found.

Rutherford, whose work laid the groundwork for the development of nuclear physics, was awarded the Nobel Prize in Chemistry in 1908. Becquerel shared the 1903 Nobel Prize in Physics with the French physicists Pierre Curie (1859–1906) and Marie Sklodowska Curie (1867–1934) for their pioneering work on radioactivity. Madame Curie became the first female professor at the famous Sorbonne in Paris, and was awarded a second Nobel Prize in 1911, this time in Chemistry, for her discovery of the elements radium and polonium, the latter named in honor of her country of birth.

Although I have pointed out to you some of those who were awarded this highest distinction in the sciences where appropriate in this story, this is hardly a complete list. I do so, almost to drop a few important names. However, as you can imagine, there are dozens, if not hundreds, of others who, though less well-known, have contributed in important ways to this story.

It is interesting that today scientists are debating the age of the Universe, that is, the time since the Big Bang. It is remarkable that we can talk about this at all. Different studies obtain numbers which are between 10 and 20 billion years, only a discrepancy by a factor of 2. Many methods are used, and just as Kelvin in his time relied on the laws of physics known then, so do we, in particular Einstein's theory of gravitation, known as General Relativity. The trouble is that some estimates of the ages of the oldest stars in the Universe suggest that these are 15 billion years old, uncomfortably close to some of the numbers obtained for the age of the Universe. We simply cannot have a Universe which is younger than its contents, and so there is a need to figure out what these numbers are telling us. This will be worked out in due course, by more precise theory and measurements. If these contradict the current theories of the Universe or stellar evolution, then too bad for our theories; they will have to be modified.

Water

Almost three-quarters (71 percent) of Earth's surface is covered by water (H_2O), a compound made of the two most common chemically active elements in the Universe: oxygen and hydrogen (helium is the second most common element but it does not react chemically – it is a "noble" gas). Our oceans can store a huge quantity of energy and so act as giant stabilizers of Earth's climate. The average depth of the oceans is 4 miles, with the total mass of water, as mentioned before, about $1/4000$ of the mass of Earth. Most of this water (97 percent) is too salty for consumption. Three-quarters of the rest is solidified at the Earth's poles or inaccessibly deep underground. Only about 0.05 percent is available for consumption, a precious supply of ground water replenished by rain, which must be conserved. This is not just idle talk. I am sure that you have heard of many places which have problems with an adequate water supply. This problem becomes more severe with passing years and increasing population and consumption.

Water is a unique compound because it is liquid at temperatures that other similar compounds like ammonia (NH_3), methane (CH_4), and hydrogen sulfide (H_2S) are gaseous. It is one of the best solvents known, providing an environment where elements and molecules in solution can easily move and react with each other. The cells of our bodies, where much of this takes place, are mostly made of water. It is one of the few substances which is less dense in solid form than in liquid form. This is why ice floats, not a trivial detail. If ice did not float, we would have no icebergs to sink the *Titanic*, but also no liquid water for the *Titanic* to navigate on. This is because ice would sink and most likely accumulate at the bottom of lakes and oceans, possibly leading in time to the freezing of all the water. As it is, ice forms at the surface and then quite effectively shields the water from the cold atmosphere, like a giant igloo, keeping the rest of the water from freezing. Life, which first developed in water, would have had a difficult time developing in ice. In frozen conditions biochemical processes do not proceed, or do so extremely slowly, which is why you can store chicken in your freezer for a long time.

Two phenomena provided the elements which compose our atmosphere and oceans. As we have seen, the primordial Earth lacked these volatile compounds which were brought to Earth in the late stages of its formation by cometary bombardment. Later, gases like oxygen, hydrogen, carbon and nitrogen, which form part of certain minerals, were also released into the atmosphere by volcanic processes.

Fire

Since about three-quarters of the surface of Earth is covered by water, a quick calculation leads to the inescapable conclusion that only one-quarter is covered by continents. Most of the Earth's surface is composed of young material since over a time of roughly 500 million years most of the surface has been recycled by erosion and tectonic processes. Consequently, it is very rare to find rocks older than 3 billion years.

Looking at a map of the world, or better still a photo from space, you might have noticed that South America's east coast seems to fit like a piece of a puzzle into Africa's west coast. Well, in fact they *are* pieces of a giant global puzzle, a realization that revolutionized geophysics. Much of what has happened to Earth over its long history is a result of a process called plate tectonics. This realization, and the development of this theory, is as important to geophysics as nucleosynthesis is to astrophysics, and the structure of DNA is to biology.

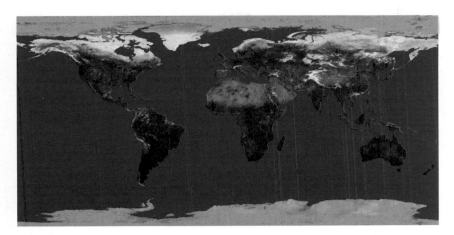

Forty percent of the Earth is usually covered by clouds. This composite image combining many cloudless views shows the Earth as it would be seen without clouds. Note how the two sides of the Atlantic Ocean fit as if they were parts of a giant puzzle. (NASA GSFC Scientific Visualization Studio)

Alfred Lothar Wegener (1880–1930), a German astronomer and meteorologist born in Berlin, like others before him, proposed this, at the time preposterous, idea in 1912. Unlike his predecessors, he based it on the evidence provided by the similarity of the rocks, fossils, and geologic structures observed on opposite coasts of the Atlantic ocean. Nevertheless, this was not sufficient for most of his contemporaries who rejected the idea because they judged the evidence circumstantial and incomplete, and more importantly because no geophysical mechanism could be identified which would plausibly be able to move the continental masses. The notion was also not in accord with the view held by most geophysicists of the time, that the continents had always been where they are.

However, starting in the 1960s, detailed studies of the ocean floors, and a knowledge of the internal constitution of our planet, obtained from studying the behavior of seismic waves as they propagated through it, produced the necessary data to prove Wegener's ideas and find a mechanism for the motion. Plates of the lithosphere float and move on top of the pliable asthenosphere, carrying along the continents and oceans. The internal heat generated by the radioactive decay of uranium, thorium, and potassium melts the rocks of the asthenosphere and drives the motion of the plates. So, in the end, the continents are gradually moved by the stored energy produced by a distant and past supernova. We are children of the stars in more than one way.

Plate tectonics was the process that created the continents and had far-reaching impact on past climate. The location and size of the continents determined the reflectivity of Earth's surface, the location of oceans and their currents, the quantity of coastal land, the amount of glaciation, and the separation between ecosystems. All this affected the evolution of life in a complex way.

Each year, a few inches of ocean crust are added at the mid oceanic ridges of the Atlantic, Pacific, and Indian oceans, pushing the plates apart and therefore moving the continents which are sitting on top of the plates. These oceanic ridges are colossal wounds at the bottom of the oceans, spanning thousands of miles on the Earth's oceanic crust, through which material from the molten interior oozes out, creating new oceanic crust. Precise modern measurements using global positioning satellites (GPS) and the radio astronomical technique of very long baseline interferometry (VLBI) confirm the movement of the plates at rates of a few inches per year – continental drift is a fact. At this slow rate, it took about 200 million years to build up the Atlantic Ocean, about one galactic year. At that time, in the Jurassic period (200 million years ago), South America was joined with Africa, and North America with Europe. This one big super-continent, which was christened Pangaea (meaning all lands) by Wegener, extended from pole to pole and contained all the land on Earth. A total of about 10 large plates have been identified, together with about a couple of dozen smaller ones. Of course, as these plates are being pushed apart at the places where new ocean crust is formed, they must collide somewhere at the other end. Here the plate with the densest material, composed of oceanic crust, plunges under the other one in what is called a subduction zone, producing volcanoes and new mountain ranges.

When the plate carrying the Indian subcontinent, moving northward, collided with Asia 40 million years ago, this violent slow-motion process resulted in the creation of the great Himalayan mountain range 10–20 million years ago. Here the collision was between two continental plates and the results were spectacular since neither plate had a propensity to sink. So far, India has penetrated about 1500 miles (2500 km) into Asia, a colossal collision. Not far away, the Arabian plate is slowly being ripped apart from the African plate at a rate of about 1 inch per year. As they separate, a new ocean is being created which will be 150 miles (250 km) wide in about 10 million years. The Red Sea and the gulfs of Suez and Aqaba mark the beginning of this event. At other places, plates slide past each other in a parallel way, generating enormous friction forces. As the plates try to move past each other, the friction force will increase until the two plates slip, instantly releasing a large amount of stored energy (like letting go of a compressed spring), generating earthquakes in the

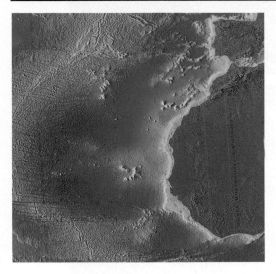

Top: the mid-Atlantic ridge is one part of a large network of under-ocean mountain ranges. It bisects the Atlantic ocean from south to north for about 9000 miles (15 000 km). At the center of the ridge is a deep valley several miles wide, a giant wound on Earth where material oozes out from the interior, generating new oceanic crust. This global ocean bathymetry map was constructed by blending depth soundings collected from ships with detailed gravity anomaly information obtained from the Geosat and ERS-1 satellite altimetry missions.

Bottom: a detailed map of a section of the mid-Atlantic ridge shows peaks that rise over 6500 feet (2000 m) above the surrounding ocean floor. Fractures which offset the general north–south pattern can be clearly seen (NOAA)

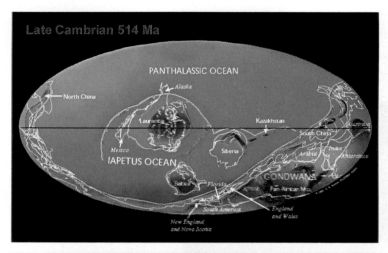

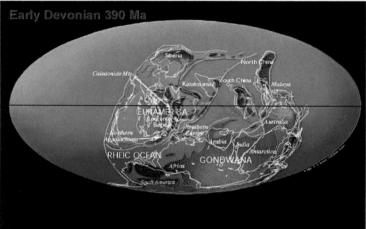

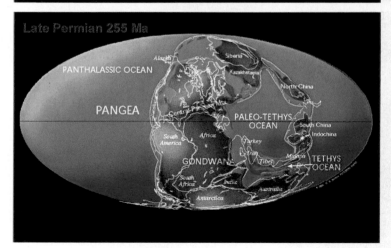

A reconstruction of the position of Earth's land masses over the last 500 million years, when Pangaea first formed and later began to break apart. The last panel shows the way the continents will be arranged 50 million years from now, if we continue present-day plate motions. The Atlantic and Indian oceans will widen, Africa will collide with Europe closing the Mediterranean, Australia will collide with South East Asia, and California will slide northward up the coast to Alaska (C.R. Scotese (www.scotese.com))

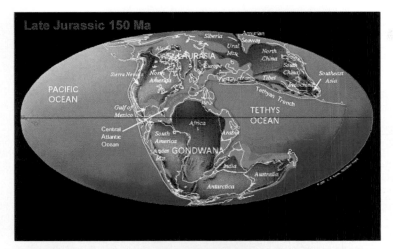

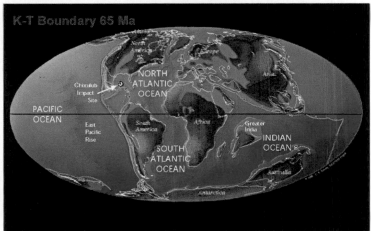

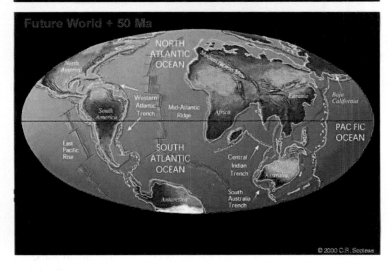

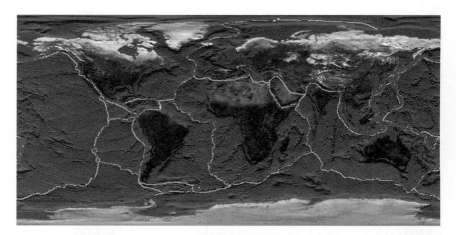

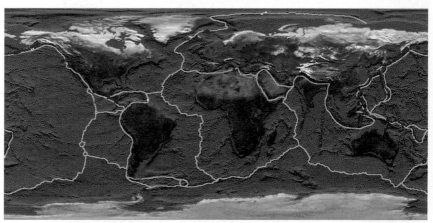

The Earth's solid surface is divided into a mosaic of tectonic plates (blue lines). As these plates move and shift relative to one another they release enormous amounts of energy in the form of earthquakes, and where oceanic plates are driven under continental plates the molten rock eventually rises to the surface, forming volcanoes. The yellow dots on the top map indicate the location of earthquakes with magnitudes greater than 4.5 that occurred between 1980 and 1995. The red triangles on the bottom map show volcanic eruptions from 1960 to 1994. It is clear that most of these are concentrated along plate boundaries. (NASA GSFC Scientific Visualization Studio)

process. Hundreds of earthquakes occur along the edges of the plates some-where on Earth every year. The famous San Andreas fault in California is at the edge of the North American plate which moves southeast at about 2 inches per year with respect to the neighboring Pacific plate.

Plate tectonics was a triumph for the uniformitarian view of geology, explaining as it does the shaping of Earth's surface as the result of a gradual process operating through enormous time spans. There was no need or reason to think of catastrophic events in the past although, as we shall see, catastrophism in a new embodiment returned to modern science.

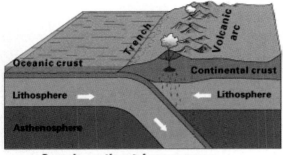

Oceanic-continental convergence

The fundamental processes of geology.

When oceanic crust collides with continental crust it will sink, because of its higher density, producing a trench where it is *subducted*. It will heat up and melt as it sinks several miles, producing earthquakes, spawning volcanoes, and creating mountain ranges. The Pacific Nazca plate pushes into and is being subducted under the South American continental plate creating the towering Cordillera de los Andes.

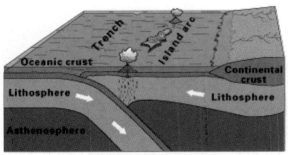

Oceanic-oceanic convergence

When two oceanic plates converge one will be subducted under the other forming a deep oceanic trench and a series of volcanos which eventually rise above the water to produce volcanic islands. The Marianas and Aleutian islands of the Pacific are such a formation.

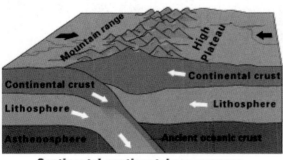

Continental-continental convergence

When two continental plates collide head-on, one will eventually be subducted and at the same time the other will be pushed upward producing mountains and high plateaus. When the Indian plate collided with the Eurasian plate it raised it over the course of millions of years creating the Himalayas and the high Tibetan Plateau behind it.

(USGS)

In September 1930, Wegener headed an expedition to resupply a meteorological station in Greenland with fifteen sledges and 4000 pounds of supplies. Although thirteen of his fourteen men turned back because of the extreme cold, he pushed on against inclement weather to reach his goal after five weeks of travel. Eager to return, he started the next morning but never made it. His frozen body was found the next summer. He never had the satisfaction of seeing his ideas proven right and accepted as such.

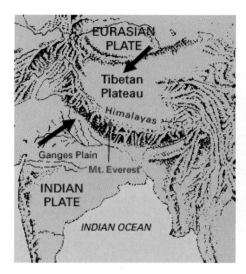

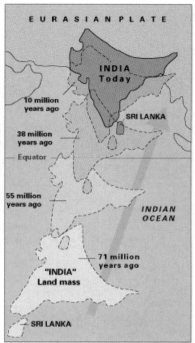

Top: about 40 million years ago the Indian plate collided with the Eurasian plate after moving northward for millions of years at a rate of about 30 feet per century. The collision gave rise to the mighty Himalayas which stretch about 1800 miles (3000 km) along the border between India and Tibet. The colossal collision that created them is still going on.

Bottom: the snow-capped Himalayas arc with many peaks rising to heights over 24 000 feet (7300 m). Toward the horizon, hazy and dusty conditions exist on the Indo-Gangetic Plain of northern India (top left) and the Takla Makan Desert of western China (top right). Numerous deep blue lakes are seen on the Tibetan Plateau, a windswept, treeless, cold and harsh place at the "top of the world," to the right of the picture. Its average elevation is about 16 000 feet (5000 m). (USGS and NASA)

Main image: the Galileo spacecraft was 300 000 miles (500 000 km) from Earth as it looked back and obtained this image of the Earth in December of 1992. The long and narrow Red Sea is prominent. It is opening up slowly as the African and Arabian plates separate. The Nile river can clearly be seen ending in a dark fertile delta on the Mediterranean coast. The inset, covering northeastern Egypt, northwestern Saudi Arabia, southern Israel and a small piece of Jordan, shows the Sinai peninsula as seen from the Space Shuttle. The Suez canal is located at the end of the Gulf of Suez on the left of the peninsula, the Gulf of Aquaba is to the right of the peninsula, and the Dead Sea can be seen to the upper right. (NASA/JPL/Caltech)

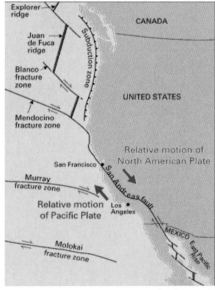

The San Andreas fault is a great fracture in the Earth's crust and the principal fault in an intricate network of faults extending more than 800 miles (1200 km) from northwest California to the Gulf of California. It is the boundary of the North American plate to the east and the Pacific plate to the west. The Pacific plate, moving northwest in relation to the North American plate, causes numerous earthquakes. This perspective view of a portion of the fault was generated using data from the Shuttle Radar Topography Mission (SRTM), which flew on NASA's Space Shuttle, and a true-color Landsat satellite image. The view shown looks southeast along the San Andreas where it cuts along the base of the mountains in the Temblor Range near Bakersfield. The fault is the distinctively linear feature to the right of the mountains. (Main image NASA GSFC Scientific Visualization Studio; map USGS)

Vital carbon

The composition of the Earth's surface, including the oceans and the atmosphere, is the result of a series of geochemical cycles which transport particular compounds between different parts of the Earth. These complex natural cycles will reach equilibrium over some time span and therefore maintain a constant proportion of compounds in the environment. Water, for example, is cycled between the atmosphere, oceans and lakes, and underground reservoirs, some of it passing through us at some point. Disturbing a cycle can, in time, lead to a different final composition, which could have quite a severe impact on life on Earth. Biological systems are an important part of these cycles, having clearly influenced, and sometimes determined, the physical and chemical nature of Earth's surface, and this has in turn affected the course of their own evolution. The biological and physical worlds are interrelated in a complex way, a fascinating dance of different biological and geophysical processes, keeping in step so as not to stumble and fall.

One of these cycles has determined the gaseous content of our atmosphere, and its evolution. The accumulation of oxygen produced by photosynthesis led to an atmosphere that was toxic to those living 1 billion years ago and required drastic changes in life forms and the development of aerobic organisms. To migrate from the seas onto land, life needed protection from ultraviolet radiation, and this was provided by the newly crated ozone layer.

Apart from its importance in relation to the greenhouse effect, carbon is the backbone of life in the sense that the chemistry of living systems is based on this element. *Life is carbon based* as is our modern industrialized economy. This is because, among all elements, carbon has a unique ability to form a variety of bonds with other elements, including other carbon atoms. This leads to the formation of a wide variety of molecules of different sizes, shapes, and properties: the bio-molecules of life.

When a carbon atom combines with two oxygen atoms to form carbon dioxide it releases energy as heat. This is the basis of the combustion that generates energy to move cars and fly airplanes, generate electricity and barbecue hamburgers. It also happens at a slower pace in the cells of our bodies to keep them warm, and to produce the energy that moves my fingers as I type these words on the keyboard, and the energy that you are using as you read this. This slower process is called respiration; it works by using barbecued hamburgers (for example) as a source of carbon, and the air you breathe as a source of oxygen. The carbon dioxide and water you produce as a byproduct of this reaction are eliminated through your lungs as you exhale. Water is also eliminated in other ways.

In passing, a diamond is just a special form of crystallized carbon, with each carbon atom strongly bonded to its neighboring carbon atoms. A diamond is not only a pretty stone, but a very useful one. It will scratch any other material, and cut it if you wish, since it is the hardest mineral known. Diamonds can only be polished and cut by other diamonds. In contrast, take the same carbon atoms and arrange them differently, so that the atomic bonds are different, and you obtain a compound which is so soft that just rubbing it on paper will waste it away: graphite. This is another useful material, which you use every time you write something with a pencil, and as a lubricant. In fact, it is known that diamonds turn spontaneously into graphite. Don't run to check your jewelry as this is a very slow process, taking longer than the age of the Earth. But, in truth, diamonds are *not* forever.

Since the time of Earth's formation the total number of atoms of a particular element has not changed, disregarding the minute quantity transformed by nuclear processes. A carbon atom, formed inside an ancient star and now in one of your brain's neurons, perhaps one that at this very moment is processing the information you read, may have been part of the wing of an Archaeopteryx (ancient wing), a small dinosaur, flying 150 million years ago. Upon its death, this animal bio-degraded and the carbon atom became part of a bacterium. At some point in history the bacterium was taken by a river to the ocean where it became embedded in sediment to form part of calcite rock. Tectonic processes then took it into the interior of the Earth where after millions of years it became part of hot molten magma. Later a volcanic eruption spewed it into the air as part of a carbon dioxide molecule. It was later incorporated, via photosynthesis, into the leaf of a spinach plant and entered your body as part of some spinach that you were force-fed when you were a child.

Carbon is constantly moving, at variable rates, between different reservoirs on land, in the ocean, and in the atmosphere – the carbon cycle. Any change in the rate of any one of these processes can change the quantity of carbon in any one of these reservoirs. The most vulnerable and important of these is the atmosphere. Carbon dioxide resides in the atmosphere with an average lifetime of several hundred years. Thus some carbon dioxide in the air today was produced as the last unhappy dodo was being barbecued on the island of Mauritius 500 years ago. Although it is only a minor component, the total amount of carbon in the atmosphere is quite large, 750 gigatons of it (in the form of CO_2), where a gigaton is a billion metric tonnes (2200 billion pounds). This corresponds to an average CO_2 concentration by volume of 360 parts per million (ppm). A similar quantity of carbon resides in vegetation, mostly in the cellulose of trees in the forests of the planet, which we are sadly eliminating at an alarming rate, and the soil contains about three times as much.

The largest reservoir of carbon on our planet, containing as much as 90 percent of all carbon, resides in carbonates of the Earth's crust. The principal minerals containing carbon are limestones of which the main constituent is calcium carbonate or calcite ($CaCO_3$). Another common mineral is dolomite ($CaMg(CO_3)_2$). These minerals formed over geologic time by reactions in the Earth's oceans which combined dissolved atmospheric carbon dioxide with calcium and magnesium originating from the weathering of rocks, which then precipitated to the ocean bottom. This process is today balanced by the return of carbon dioxide to the atmosphere by volcanism as these minerals are heated inside the Earth. In this way, together with photosynthesis, the initially high concentration of carbon dioxide in the atmosphere was removed over billions of years.

Deep in the crust, fossil fuels – coal, natural gas, and oil – exist in reservoirs which contain about 5000 gigatons of carbon (80 percent of this in coal), about seven times the amount in the atmosphere. They are called fossil fuels because they are the product of former life. Photosynthesis in plants produces molecules containing carbon – so-called organic molecules – from water and atmospheric carbon dioxide, releasing oxygen in the process. The burial of these molecules removed carbon "permanently" from the atmosphere, the process which led to the current atmospheric composition, as we have seen. When these fossil fuels are burned to obtain the energy stored in them, they release carbon dioxide back into the atmosphere. Were we to burn a large fraction of this, it would increase the atmospheric carbon dioxide by a large factor and this would remain in the atmosphere for a long time. This is the crux of the problem, the problem of global warming, and our future on this planet. Surface temperatures could increase by several degrees as a consequence of the relatively rapid addition of substantial quantities of greenhouse gases to the atmosphere, with disastrous consequences for life on Earth. I will return to this line of thought in the final chapter.

Ancient Earth

You would have hardly recognized our planet had you visited there about 3 billion years ago, in the so-called Archean (from the Greek *archaios*, "ancient") eon. You would have had to use a space suit to survive in this alien world. The area of the continental surface was then much less than today,

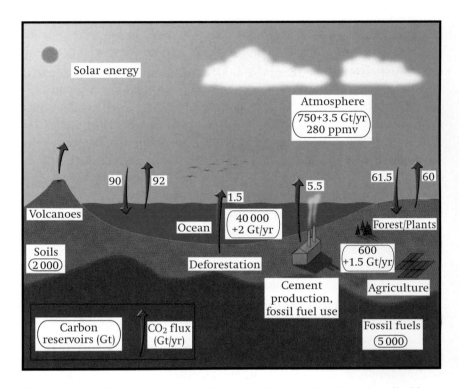

The carbon cycle. Since the beginning of the industrial revolution the carbon dioxide concentration in the atmosphere and in the oceans has increased at an accelerating pace. Deforestation has at the same time reduced the rate at which carbon dioxide is removed form the atmosphere. A complex series of biogeochemical cycles, with various sources and sinks of carbon dioxide, must be studied and fully understood to give a better model of the future of the climate. Although some carbon reservoirs are very large, they are quite stable and do not play a role in rapid climate change. The Earth has about 100 million Gt (gigatons) of carbon, 80 percent of it locked in sedimentary rock. The rest is dissolved in the oceans, is in the atmosphere, in fossil fuels, in soil, and in plants. What is important is the rate at which carbon moves from one reservoir to another. In particular, fossil fuel combustion and cement production, together with deforestation, produce 7 Gt per year. One half of this remains in the atmosphere and increases its CO_2 concentration while the rest is reabsorbed in the oceans and in the biosphere. (José F. Salgado)

since the processes which build continental crust, the surfacing of molten rock from the interior of the Earth, were only just beginning. The land surface under the hot, dense, original atmosphere was barren, the Sun looked red, and the sky appeared light orange, rather than blue. Plants and animals did not exist yet, and in any case would not have survived in this environment, with no ozone to protect them from ultraviolet radiation, and no oxygen to breathe. This is why you would have needed a space suit.

Cosmic projectiles had ceased hitting the Earth at a high rate but such collisions still occurred every few thousand years, creating great upheavals. The scars of these would have been visible all over the landscape, mixed in with volcanic calderas; erosion had not had enough time to erase the evidence. Volcanic activity would have been vigorous, driven by the great internal heat which kept a large reservoir of molten rock (magma) under the surface. The continents and the oceans were still hot, the temperature at the surface reaching the boiling point of water, although water did not boil because the atmospheric pressure was several times that of today. The large quantity of carbon dioxide in the atmosphere kept it that way, even though the Sun was less bright. Atmospheric water vapor, supplied by volcanos and comets, condensed and fell as torrential rain. The rain was acidic because carbon dioxide combined with water to form carbonic acid, a sort of "Perrier" rain. Over the eons this intense rain, which formed the oceans, also removed a good part of the initial carbon in the atmosphere.

Looking at the night sky from this ancient world, you would have seen only the brighter stars and planets through the foggy atmosphere. However, even if you had been able to see all the stars, it would have been a sky which both you and Tycho Brahe would not have recognized. You would be looking at a completely different part of the Milky Way because between that time and today our Sun, traveling once around our galaxy every 250 million years, would be at a totally different place, surrounded by other stars. The Moon, recently formed, reddish because of the thick terrestrial atmosphere, would have been much closer than today, looking many times larger and brighter than today, an awesome view. Because it was closer, the Moon would have raised gigantic tides, tens of times higher than today. Ocean water would have been pulled onto land for quite a long way, by the tidal attraction of the Moon, leaving pools of water which got replenished every few hours since the Earth then rotated much faster than today. Looking at some of the more protected shallow pools of water which had formed on the landscape, you would have observed slimy roundish structures a few feet in diameter. Soft, and with a greenish-bluish tint, they were covered by layers of sand brought in by the tides, which protected them from ultraviolet radiation. You would recognize them as stromatolites. *You would have discovered life on Earth!*

Life. Of the uncountable processes taking place in the Universe, life is surely the most important one, or at least and understandably, so it seems to us. Let us take a short detour and look at life and its evolution on Earth in the next chapter.

Chapter 5

Life

Life's bad enough as it is, without wanting to invent any more of it.[1]

The myriad organisms we observe, what we call biodiversity, are seen to be in essence the same basic thing. Wherever we look we observe the same biochemistry and the same ingredients. It all points to the fundamental oneness of life, to the astonishing idea that all life has descended from a common origin in the distant past, but the way it arose still escapes us. If you have seen the blueprints for a nuclear power plant or a jumbo jet you know how complex they are. Architects and engineers worked long hours to produce them. Yet a 747 is simplicity itself when compared with even a small animal composed of hundreds of billions of cells carefully organized into a living system. Who are the architects and engineers of life? Where is the million page blueprint?

Biosphere

Earth's biosphere, the region where life is found, is a relatively thin shell a few miles in thickness surrounding our planet. This habitable zone of Earth, like that surrounding a star, is in great measure determined by the temperature range over which water is liquid. At heights of a few miles above the surface it becomes so cold that water freezes. Going in the opposite direction, the temperature of the crust increases, reaching the boiling point of water 2 miles beneath the surface. If the Earth were the size of a basketball, all life, everything that happens between animals and plants, all the activity of our cities, all the humongous traffic jams, all the havoc we create with our constant quarrels and cruel wars, all the love scenes, all the rock concerts, and all the football games take place inside a layer on the skin of this basketball thinner than the width of a hair.

---

[1] Douglas Adams, *The Restaurant at the End of the Universe*, Six Stories, p. 248.

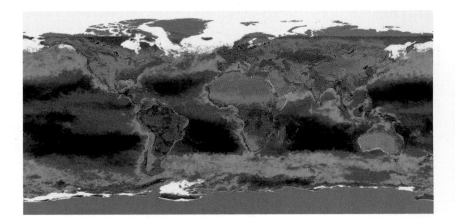

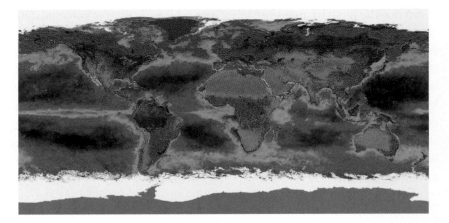

Top: winter in the north. Bottom: summer in the north.

Plants on land and in the ocean (phytoplankton – microscopic plants that grow in the upper sunlit reaches of the ocean) form the basis of the food chain for most life. They contain chlorophyll, a green pigment that they use during photosynthesis. Using satellite-based sensors, we can measure chlorophyll concentrations on land as well as in oceans, lakes, and seas to indicate the distribution and abundance of vegetation. Since most animal life relies on vegetation for nutrition, directly or indirectly, these images represent a snapshot of Earth's biosphere. Red and orange colors in the oceans denote high concentrations of phytoplankton, seen mostly in coastal areas with high concentrations of nutrients. Yellow and green denote moderate concentrations, while turquoise and blue denote the lowest concentrations. On land the rain forests are shown in dark green, while lighter green shows subtropical forests, farmlands, and some dryer areas. The great desert areas of the world, including areas covered in snow and ice, with little or no plant growth are shown in brown and yellow. By studying maps such as this one over time we will be able to understand and better predict the effects of human activity and natural changes on the biosphere. The seasonal changes in the biosphere can be clearly appreciated in these two images separated in time by six months. (Image provided by Orbimage)

The biosphere is populated by a bewildering variety of organisms, several million different species, from tiny invisible bacteria to large trees and animals like us and the cheetahs. This great diversity of organisms occupies every part of the biosphere, and surprisingly also places where you would not expect to find it. Some forms of life, so-called autotrophs (from the Greek *auto*, "self," and *trophos*, "feeder") such as plants, use elements from the environment, and energy extracted from sunlight or from chemical reactions, to build the molecules they need. Other forms of life, such as animals, which are heterotrophes (from the Greek *hetero*, "other"), use the energy stored in the molecules made by others for their metabolism. Autotrophs and heterotrophes live in a complex interdependent cycle composed of those who eat and those who are eaten, the predators and the prey, the food chain of the ecosystem which is the biosphere. The more we study the intricate web of life, the more interconnected it becomes, many forms of life dependent on many others in ways we do not fully understand. These dependencies form the basis of the functioning of the biosphere, and this is one of the reasons why we worry about the loss of biodiversity.

We are produced by the union of two tiny cells, we spend a lifetime eating other life forms to produce the energy and the materials we need to maintain our organism, only to die and serve as food for other tiny life forms. Every year about 100 million persons die, about 275 000 per day, three per second, to be bio-degraded by the work of bacteria. Quite depressing isn't it? At the same time about 200 million persons are born every year (six per second!), and this is leading to a little problem.

To understand if the Universe might be teeming with life, and in particular advanced life, we need to learn as much as possible about the origin of life and its evolution on Earth. There are three questions related to the origin of life whose answers we would like to know: *when*, *where*, and *how* life appeared. The answer that we used to give to these questions was that life arose all the time, in any corner of a dirty kitchen, by the process of spontaneous generation in decomposing matter. This was believed until Louis Pasteur (1822–1895), an eminent French chemist and microbiologist, proved that this was not so, that spontaneous generation did not occur if a system was properly isolated from outside contamination.

Life

Before considering life on Earth and its evolution, let us briefly describe what life is. It feels as if we intuitively know but further reflection leads us to conclude that we really don't. It is possible to go into great lengths discussing this, but I will state only the fundamental properties that we can all agree are part of life. Life is a system able to replicate and assemble itself using energy obtained from chemical processes or light. Living systems do not need outside information for their maintenance and reproduction. An essential feature of life is that it is not perfect (as if you hadn't noticed). In relation to this story, however, it is replication of the genetic information that is *not perfect*, errors are made, paving the way for the process of evolution, through changes which accumulate over many generations to produce new forms of life. Compared with inanimate matter, which is composed of a random mixture of simple chemical compounds, life is infinitely more complex and, what is more important, highly organized.

An important development toward the understanding of life was the realization that the cell is the structural and functional unit of a living organism. Everything alive is made of cells, one or billions of them. This was a view that transformed the field of biology and was advanced by the influential German physician and anthropologist Rudolph Virchow (1821–1902). Life comes only from other life through cell division, except of course at its origin. The two tiny cells which started us divided and multiplied to form the great variety of cells, about 200 different ones, in our bodies. These cells have different structures and roles, such as the red and white blood cells, the neurons of the brain which at this instant are helping you understand these symbols on the paper, the cells in our skin or the special reproductive cells. All these cells perform specific functions for the complex organism they form, following a carefully orchestrated plan. Cells continuously generate energy (otherwise you'd be cold), produce proteins, fight intruders, and transport oxygen or chemical signals. Cells divide, multiply, and change in form and function, as living things develop and grow and as a need for more cells arises. We are a huge bunch of cells, about 10 000 billion of them, organized into an unbelievably complex system that is continuously renewing itself, until it dies. Our brain, a special organ well cushioned inside our head, contains as many neurons as there are stars in our galaxy. It is the source of our consciousness and of our thoughts, which seem to transcend our physical existence, and makes us unique in the animal world.

Life consists of three major domains, the *bacteria*, *archaea*, and *eukarya*, with viruses as a fourth separate group which, because they cannot replicate on their own, are not considered to be really alive. Unicellular bacteria and archaea are Earth's earliest known inhabitants, and are the ancestors of our ancestors. They are simple microscopic organisms measuring a few tenthousandths of an inch, this being about 100 times larger than a virus. They are made of just one prokaryotic (from the Greek *karion*, "kernel," and *pro*, "before") cell with a rigid outer wall, no internal membranes, and no nucleus. These microorganisms occupy a wide range of habitats and can survive where other life forms would not even dream of going for a short visit. They represent the most successful inhabitants of this planet, and I suspect will be around long after our species disappears. Although invisible to the naked eye, they outweigh visible life by at least a factor of 10. There are many more bacteria in your gut than there are humans on Earth, currently estimated at more than 6 billion persons. Prokaryote cell division is fast and simple: a cell is able to divide into two "daughter cells" in a few minutes. If a cell divides in half an hour, say, you will have four cells after one hour, and sixteen after another hour. After five hours, ten divisions, you will have 1024 cells and after just five more hours you will have reached 1 048 576 cells, five hours after this you will have reached an incredible number (I will let you figure this one out). So you see how bacteria can increase their numbers prodigiously in a short time, a consequence of what is called exponential growth, as happens during a bacterial infection or when you make yoghurt. If the process were to run unchecked for just two days, it would cover the entire surface of our planet (oceans included) with bacteria up to your ankles. Of course this does not happen, since they soon run out of nutrients or habitat.

Many archaea and some bacteria are extremophiles, which means they are able to live and thrive in extreme conditions such as high temperature, high acidity, high salinity, and cold. Most are anaerobic, meaning that they live in conditions without oxygen, presumably resembling the conditions on our primitive Earth, and maybe also conditions on other bodies of the solar system. The Río Tinto, near Huelva in Spain, is so named because its water is the color of *vino tinto*. Although you would gladly drink its namesake, you probably would not want even to wash your hands in the river, since it has a high concentration of heavy metals and sulfuric acid. However, there are bacteria living happily in this river that are capable of oxidizing sulfur and iron, giving the Río Tinto its red hue. (Well, I have not actually asked them if they are happy.)

Studies of samples of material drilled from about 2 miles deep into the Earth's crust have revealed anaerobic bacteria that live at the high temperatures prevailing there. These bacteria derive energy from hydrogen gas, and nutrients from inorganic chemicals. At the deep mid-oceanic ridges, where new oceanic crust is generated and continents separate, material from deep inside the Earth's mantle is released, forming what are called hydrothermal (hot water) vents. Scientists studying these vents were greatly surprised in 1977 when they discovered an exotic ecosystem thriving in the total darkness of the deep ocean. Here, far from the universal energy source for life, our Sun's light, hydrogen sulfide (H_2S)-oxidizing bacteria live at near boiling

Through chimney-like vents deep in the oceans, hot mineral-rich water at a high temperature rushes out to meet the cold ocean water, and the dissolved minerals solidify, giving the appearance of smoke. Scientist exploring these vents were surprised to find life thriving in these environments.

Here, far from our Sun's light, the universal energy source for life, hydrogen sulfide (H_2S)-oxidizing bacteria live at near boiling temperatures. These organisms use chemical energy (chemosynthesis), rather than photosynthesis, for their metabolic processes. They are the bottom of a food chain which allows larger organisms such as the tube worms seen here, large clams, mussels, and crabs to feed by way of other bacteria which live symbiotically with them. (OAR/National Undersea Research Program/NOAA)

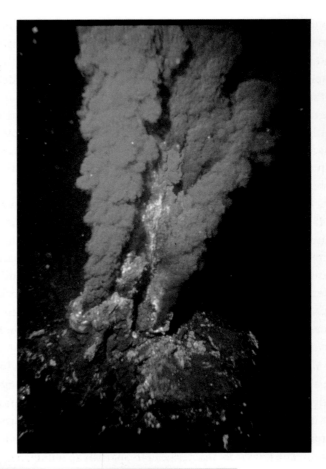

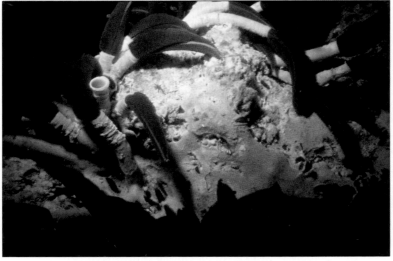

temperatures. So life can be found in places where we would not have thought it could possibly exist, a warning about our predictions concerning life elsewhere in our solar system.

The eukarya include those life forms of the three kingdoms with which we are most familiar: animals, plants, and fungi, which are built up of a large number of eukaryotic (from the Greek *eu*, "good") cells. The so-called protists are mostly single-celled microscopic eukarya like protozoa and algae. The eukaryotic cell has about 10 000 times the volume of, and is quite a bit more complex than, the prokaryotic one. In the central nucleus are the chromosomes (from the Greek *chromo*, "color," and *soma*, "body") containing genetic information. Chromosomes are so called because they can be stained with special dyes which make them stand out in bright colors under a microscope.

Many different organelles, tiny structures bound by their own membranes, are found in the cytoplasm of the eukaryotic cell, serving special functions. In the cells of plants, photosynthesis takes place in an organelle called a chloroplast. The mitochondria are organelles where oxygen is used to break down molecules and derive energy. Mitochondria are similar to the group of purple bacteria, which includes *Escherichia coli*, and the chloroplasts are similar to cyanobacteria. Careful studies of their properties has led to the exciting proposition that, at some point in the early history of life, a primitive protist with its soft membrane, engulfed a bacterium and this marked the beginning of a mutually beneficial relationship. Alternatively, a bacterium invaded another one, but it became advantageous for the predator not to kill its prey, and so a cooperative venture was started, a symbiosis – a mutually beneficial relationship – that has lasted for over 2 billion years. More generally, without bacteria living symbiotically with us, we would not survive.

History

The history of life on Earth is divided into several intervals defined by the *geological time scale*. Two eons, the long Precambrian and the shorter Phanerozoic (from the Greek *phaneros*, "visible," and *zoe*, "life") are subdivided into eras, which are subdivided into periods, and then into epochs. The first multicellular organisms, called metazoans, appear in the fossil record about 550 million years ago at the start of the Paleozoic (ancient life) era, which is at the beginning of the Phanerozoic eon. For a very long time before this, life was microscopic. Small dinosaurs and mammals first appeared 250 million years ago, early in the Triassic period of the Mesozoic (middle life)

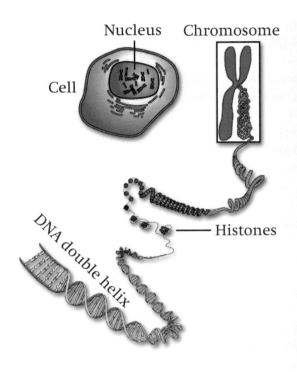

Nucleus

Chromosome

Cell

Histones

DNA double helix

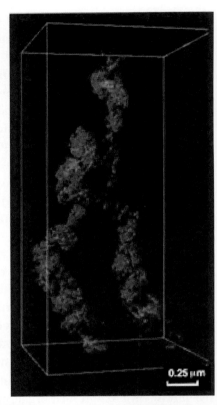

0.25 µm

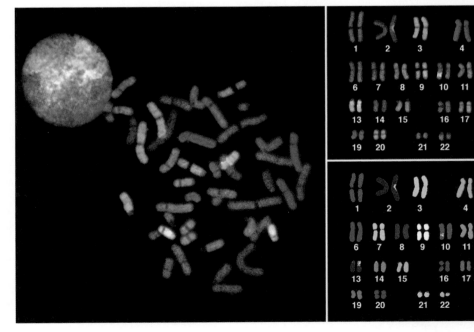

era. They lived in Pangea. The boundaries between different intervals of the geologic time scale are not defined by a particular number of years having elapsed, but by places in the fossil record that show a significant change, so the geological time scale is also a record of the history of life.

Fossils are the key to understanding past life on Earth, and its evolution, as they show life forms that are no longer living, which can be placed, after careful anatomical study, into some category from kingdom to species, relating them to current life forms. A species is composed of a population of organisms that can interbreed naturally, and extinction of a species means that this type of plant or animal will never be seen again. *Diamonds are not forever but extinction is.* From the fossil record, Cuvier had established the important fact that there were life forms in the past that no longer exist, and there are present life forms that did not exist in the past. There was a time when there were no birds and no humans. Birds and humans must have developed from non-birds and non-humans. Since all life comes from previous life (except at the origin), it must have evolved – there is no question about this. The fossil record contradicted the idea that all forms of life had always existed, the descendants of an original pair preserved in Noah's ark.

The fossil record also shows instances where transitional forms of animals, between an earlier and a later one, can be recognized. Possibly the best-known case is that of Archaeopteryx, a fossil primitive reptile, the size of a pigeon, which flew about 150 million years ago. The first example was found in 1861 in a Jurassic limestone formation in Bavaria, Germany. A few more have been found since then. These fossils contain clear impressions of feathers on the arms and tail, the first feathered creature ever found. At the same time the fossil has features characteristic of a maniraptor, a small dinosaur with a body plan resembling an ostrich with arms and clawed hands instead of wings. Modern birds are the descendants of this flying dinosaur so, in this sense, it did not really become extinct.

Inside the nucleus of each cell of your body (about 100 trillion of them) 23 pairs of chromosomes contain the DNA that stores the genetic information (the genome) necessary to build a human being. The structure of individual chromosomes can be studied by electron microscopy tomography, a technique which allows a reconstruction of a chromosome's three-dimensional structure, as shown at top right.

Several techniques are used to stain and visualize the chromosomes for their study. The color image of the human karyotype (the full set of chromosomes) was produced by the technique of spectral karyotyping, whereby different fluorescent molecules are introduced into specific regions of the DNA of each chromosome, painting each with a different color. (Top left and bottom National Human Genome Research Institute; top right courtesy Peter Engelhardt, University of Helsinki, Finland)

GEOLOGICAL TIMESCALE

EON	ERA	PERIODS/EPOCHS	EVENTS	Mya non-linear scale
Phanerozoic *(visible life)*	**Cenozoic** *(recent life)* (65 Mya to today)	**Quaternary** (1.8 Mya–today) **Holocene** (11,000 years– today) **Pleistocene** (1.8 Mya–11,000 yrs) **Tertiary** (65–1.8 Mya) **Pliocene** (5–1.8 Mya) **Miocene** (23–5 Mya) **Oligocene** (38–23 Mya) **Eocene** (54–38 Mya) **Paleocene** (65–54 Mya)	species *Homo* mammals	
	Mesozoic *(middle life)* (245–65 Mya)	**Cretaceous** (146–65 Mya) **Jurassic** (208–146 Mya) **Traissic** (245–208 Mya)	*mass extinction* (65) reptiles Pangea break-up *mass extinction* (208) first dinosaurs *mass extinction* (245)	65 245
	Paleozoic *(ancient life)* (544–245 Mya)	**Permian** (286–245 Mya) **Carboniferous** (360–286 Mya) **Devonian** (410–360 Mya) **Silurian** (440–410 Mya) **Ordovician** (505–440 Mya) **Cambrian** (544–505 Mya)	*mass extinction* (367) land animals land plants *mass extinction* (440) Cambrian explosion atmospheric oxygen high snowball Earth	550
Precambrian	**Proterozoic** (2500–544 Mya)	**Neoproterozoic** (900–544 Mya) **Mesoproterozoic** (1600–900 Mya) **Paleoproterozoic** (2500–1600 Mya)	oldest eukaryotic fossils snowball Earth	2500
	Archaean (3800–2500 Mya)		Bacteria and Archaea	
	Hadean (4500–3800 Mya)		oldest Stromatolites first life Moon formed Earth bombarded	3800

Until the discovery of *Archeopteryx lithographica* (Late Jurassic, 150 million years ago) in 1861, there were no known transitional fossils between reptiles and birds. This find showed clearly that the two groups were in fact related. Several specimens of this fossil show reptilian and avian features. The skull and skeleton are basically reptilian, with a full set of teeth and three claws on the wing, while bird traits are limited to the wishbone, where flight muscles attach, and the unmistakable imprint of feathers. (Painting by Mineo Shiraishi, fossil courtesy of Naturmuseum Senckenberg, Frankfurt am Main, Germany: senkenberg.uni-frankfurt.de)

Fossils form when hard parts of organisms, such as teeth, shells, or bones, are preserved intact, or modified in chemical composition but still preserving the original form. Dissolved minerals will often fill the small spaces in the remains of an organism and crystallize, so the fossil becomes a rock with the shape of the animal or plant imprinted in it. A different kind of fossil is produced by the remains of animals and plants which died hundreds of millions of years ago, and were buried and transformed by chemical and physical processes. We can find these remains deep inside the Earth's crust and we mine them to use their stored energy originally provided by our star, the Sun. Coal, produced from the remains of ancient plants, and oil and natural gas, substances derived from ancient organisms, are *fossil fuels*. On rare occasions, the soft parts of an organism are preserved, if they are buried so that they do not decay, or they are frozen in the permafrost of Alaska or Siberia. Sometimes the remains of animals and plants are found exquisitely preserved in amber, trapped in the sap of ancient trees.

Occasionally, only the traces of animals are preserved, such as the

Million-year-old insects are sometimes found exquisitely preserved in amber, a form of tree resin. This is a specimen of the extinct 20-million-year-old termite *Mastotermes electrodominicus*. It is possible to extract and study DNA from these insects and in this way study their relation to modern insects. (D. Grimaldi)

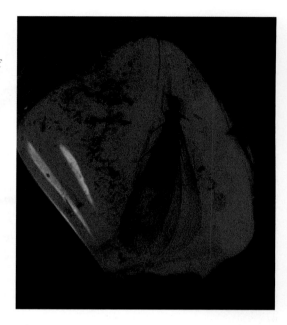

famous footprints in solidified deposits of volcanic ash, discovered in Laetoli, northern Tanzania, in 1974, by a team led by the well-known anthropologist Mary D. Leakey (1903–1996). These footprints bring us news from three hominids going for a walk about 3.6 million years ago. They provide us with a unique glimpse of these ancestors of ours, and tell us that these early hominids walked upright, long before the use of stone tools and the development of large brains. They are as significant as the trace left by those few neutrinos caught after 160 000 years of travel bringing us news about the death of a distant giant star.

We do not know where these hominids were going, but in our imagination we can follow their footprints until they join the ones we recently left on the surface of the Moon. We have come a very long way indeed.

We know that the fossil record gives us only a coarse and incomplete sample of life through the ages, a record further marred by geophysical processes, which will erase it, especially for very early times. There must have been many animals and plants that have left no trace, or of which the traces are so rare that we have not yet found them. Nevertheless, the fossil record gives us a fascinating view of the history of life on Earth, and has allowed paleontologists to reconstruct its major events. The fossil record shows extinctions of many species during a geologically short time. Such is the case at the Cretaceous–Tertiary boundary, where it is estimated that up to 75 percent of all plants and animals, including the dinosaurs, became

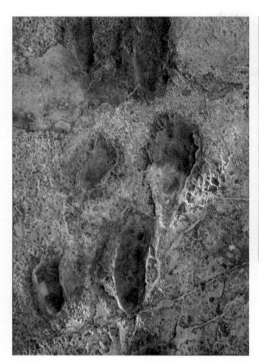

These footprints symbolize our past and our future. The barefoot ones of the past mark
the beginning of the long path that led to us. They were left some 3 600 000 years ago in
Laetoli, Tanzania, by hominids going for a walk, toward a place we do not know. They
were discovered by anthropologist Mary Leakey in 1974, preserved in hardened volcanic
ash deposited from a nearby volcano, and are the most ancient traces yet found of our
ancestors. The footprint of a boot on the moon, a technological footprint, was left there
on July 20, 1969, by Apollo astronaut Neil Armstrong: a "giant leap for mankind," toward
a place we do not yet discern. (Left courtesy of Heinz Rüther, Department of Geomatics,
University of Cape Town; right NASA)

extinct, never to appear again. At least five such episodes, known as mass
extinctions, are recorded in the fossil record. The most devastating of these
occurred at the Permian–Triassic boundary, ending the Paleozoic era 245
million years ago. As many as 95 percent of all species then living disappeared, a great catastrophe which destroyed nearly all life. The fossil record
tells us that it took about 100 million years for life to recover from this blow.
Except for one case discussed later, the particular cause of these mass extinctions is not entirely understood, but is most likely to be related to significant
environmental changes.

   The extinction of a species, just like the death of individuals, is a fact of
life occurring all the time. Millions of species have died out during Earth's

history, in fact most of the species which ever lived are extinct, and many more are dying out as you read this. I mentioned the famous dodo (*Didus ineptus*) in the previous chapter, a flightless bird bigger than a turkey and described by Dutch sailors who visited the island of Mauritius in the sixteenth century. This nice bird was extinct by 1681 and will never be seen alive again. Its scientific name clearly represents our opinion, and excuse, for why this bird became extinct. Today, a large number of species are in danger of extinction, most of them because of *us* extinguishing them, not something to be proud of. We are not ourselves an endangered species as yet, but this is not for the lack of trying. Ten thousand years ago, we eliminated the mammoths, and right now we are wiping out an alarmingly large number of animals and plants. At the start of the twentieth century there were about 100 000 tigers on the Asian continent. In only 100 years we have decimated this population so that there are less than 5000 left. Even our nearest relatives, the great gorillas and good-natured chimpanzees, are not

The world's fastest cat, the beautiful cheetah (*Acinonyx jubatus*), can reach the astonishing speed of 70 miles per hour (110 km/h) in a few seconds. It is endangered due to loss of habitat, decline of prey species and conflict with livestock farming. Cheetahs were once found throughout Asia and Africa. It is estimated that over 100 000 were alive in 1900. Today, less than 10 000 remain, most of these in southern Africa, the largest population being found in Namibia. This mother with her cub was photographed near Kruger National Park in South Africa. (José L. Alonso)

doing well. It will be a sad day on Earth when the last one dies. If, as some claim, there is a universal court where humans will be eventually judged, it will be *this* that will convict us. We will, of course, plead innocence by reason of insanity, and will be condemned to psychiatric prison where we will feel at home.

We can relate the fossil record to a time sequence because it shows up in sedimentary rocks. As the name implies, these rocks are formed by the gradual deposition of loose particles eroded from Earth's surface. They are carried by wind and water to their final resting places where they accumulate and are compacted, eventually to be transformed into rock. If you look at the face of a cliff, you can often see layers of different colors, the result of millions of years of this process and the eventual lifting of the entire structure to where you can see it, by the geological forces associated with plate tectonics. It is reasonable to suppose that the top layers were deposited after the bottom layers. The different colors are a consequence of the changing environment which changes the chemical composition of the deposits. Under the right conditions, any animal or plant caught in these deposits will leave a fossil trace. In this way, the geological and fossil records give us a sequence of events. Radioactive dating of the rocks associated with a fossil, or a direct measurement of the time since death using radiocarbon dating, gives us the actual age. This is how we can state that the dinosaurs became extinct 65 million years ago, or that a fossil stromatolite is 3.5 billion years old.

Clearly, life must have originated sometime after the formation of the Earth, unless we think of it as having arisen elsewhere in the Universe and then being transported, by some means, to our planet. This latter option is certainly possible, although something of a cop-out. We have seen that the Moon was the result of a devastating impact that melted and sterilized the Earth, so any life existing before this event would not have survived. We also know that even after this, and until about 3.8 billion years ago, our planet was probably not in a condition to sustain any life which might have formed, because of the constant bombardment of its surface. In addition, this bombardment was needed to provide Earth with the biogenic elements, and water, necessary for life. It is therefore reasonable to say that life arose *more recently than* 3.8 billion years ago.

The oldest reliably identified fossils come from rare rocks 3.5 billion years old, found in Warrawoona, western Australia. They contain finely laminated structures identified as fossil stromatolites. These are formed by the slow upward growth of bacterial communities which form mats, still found today in some shallow seas. Cyanobacteria populate the top of the mat, using sunlight to drive the formation of organic molecules, releasing oxygen in

Geologic processes will at some places expose a cut of the crust showing different layers corresponding to different geologic times. This photograph, a one-minute exposure by moonlight, shows comet Hale-Bopp at the horizon as a bonus. The Grand Falls, or Chocolate Niagara Falls, are taller than Niagara. They are located on Native American sacred land in northern Arizona. (Dennis Young)

Stromatolites (from the Greek *stromatos*, "layer," and *lithos*, "rock") look like layered rock but it is their biological origin that makes them special. They come in many different varieties and shapes and are found fossilized, or still forming today as the result of the growth of bacterial mats. The structures are built layer by layer as the top layer of cyanobacteria grows. Under the top layer other bacteria thrive, feeding from the detritus that trickles down from the top of the mat. The living stromatolites are from Shark Bay, Australia. The Petrified Sea Garden is a fossilized stromatolite ocean-reef from 500 million years ago, a time when the land that is now Saratoga Springs in New York was at the shore of a warm tropical sea. (Top Helen Stephenson; bottom Joseph Deuel)

Photosynthesis and respiration.

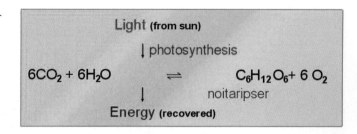

the process. Nearby, in the Apex chert rocks dated to be 3.465 billion years old, fossils of cellular cyanobacterium-like filaments have been found, the oldest ever. *We can conclude that life started on Earth surprisingly early in its history, in the relatively short 300 million year interval between 3.8 and 3.5 billion years ago.*

Starlight for life

A significant fraction of biological systems derive energy efficiently from sunlight, which, as we have seen, is distributed over a range of wavelengths centered about those visible. Each wavelength corresponds to a certain energy content, with shorter wavelengths carrying more energy. Thus ultra-violet light, with wavelengths shorter than violet light, carries enough energy to break the bonds that hold atoms together in molecules, therefore producing considerable damage to life. This is why the absorption of ultra-violet light by stratospheric ozone is so important to life.

Photosynthesis is, without question, the most important biological process on Earth, because it uses the energy provided by our star, the Sun, and produces a stored form of energy used by most living systems, directly in plants (autotrophs) and indirectly for animals that eat plants or other animals (heterotrophs). This process also stores solar energy for later use, as organic matter is buried in sediments (fossil fuels). It is also the source of atmospheric oxygen needed by modern animal life. Photosynthesis uses atmospheric carbon dioxide and water to produce carbohydrates for plant growth. Carbohydrates are molecules like cellulose, a polysaccharide – a molecule made by a chain of simple sugars, containing multiples of the unit CH_2O – which is the main structural component of trees. Photosynthesis uses energy from light and releases one molecule of oxygen for each atom of carbon used to build the carbohydrate. A large fraction of the carbon in the original atmosphere of the Earth (in carbon dioxide) was used to build carbohydrates in growing plants, and then buried in the Earth's crust.

Animals use respiration to produce energy, a process that is the reverse of photosynthesis with water and carbon dioxide as a byproduct. There is therefore a balance between the production of carbon dioxide by respiration and the burning of fuels, and its consumption by plants. Every year, about 60 gigatons of atmospheric carbon dioxide are cycled in this way. As stated before, this remains a balance as long as one side of the process is not modified drastically.

The greenish-bluish cyanobacteria (from the Greek *kyanos*, "blue") are extraordinary organisms to whom, in some sense, we owe our existence. They are among the most versatile organisms around, being able to live almost anywhere, spanning extremes of temperatures, light intensity, and acidity, even resisting doses of radiation which would kill other organisms.

(a)
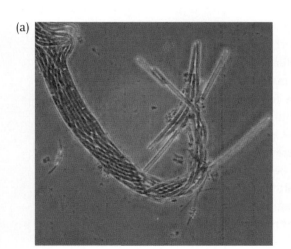

(b)
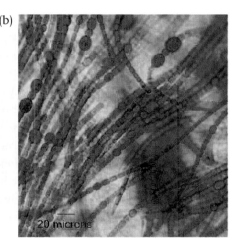

(c)
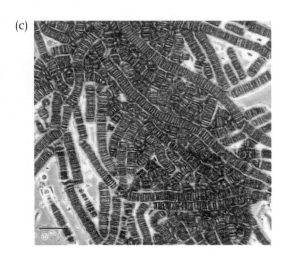

The aquatic and photosynthetic cyanobacteria can be considered as "living fossils," having changed very little over the eons. They are extremely versatile, living in different and sometimes extreme environments, and having survived all mass extinctions. Cyanobacteria are the builders of our oxygen-rich atmosphere. They are microscopic unicellular procaryotes which often grow in filaments or colonies forming bacterial mats and stromatolites.
(a) Strings of cyanobacteria from the green layer of a microbial mat from the Great Sippewissett Saltmarsh, Falmouth, MA
(b) *Anabaena spherica*
(c) *Lyngbya sp*
(Cyanosite *www-cyanosite.bio.purdue.edu/*)

This versatility has allowed them to survive with little change over geologi-cal time. They ruled our planet for billions of years while many other species appeared and, after a few million years, disappeared. The cyanobacteria sur-vived all sorts of environmental changes, and produced the most important environmental change of all: our oxygen-rich atmosphere. By the early Cambrian, about 550 million years ago, the supply of free oxygen in the atmosphere had reached about current levels.

Unity

Life presents an amazing variety of external forms, and even the simplest organisms are incredibly complex and highly organized systems. However, the design similarities between a whale, a human, and a rat are much greater than the differences, although their appearance might suggest oth-erwise. (Sometimes we even refer to some humans as "rats" but this is not a consequence of any profound insight.) True, all these are mammals, using essentially the same processes for respiration, food processing, and envi-ronmental sensing. But even a human and an orange tree, although very dif-ferent in the way they breathe, process food, and sense the environment, have similar design and processes at a deeper cellular level.

Looked at from a different perspective, all life forms are in essence the same basic thing. They are built with the same ingredients, those biogenic elements which, not surprisingly, are the most abundant in the cosmos thanks to the stars. Ninety eight percent of all organic matter is made out of just four elements: hydrogen, oxygen, nitrogen, and carbon. Although helium is the second most abundant element in the Universe, it does not react chemically, and so is not a component of any molecule. About three-quarters of living matter is made of water (hydrogen and oxygen), which is why it cooks well in a microwave oven. Over one half of the rest is carbon, and the rest is calcium, phosphorus, potassium, sulfur, chlorine and traces of some other elements. Except for nitrogen and carbon, resident in our atmosphere, these elements are also the most common ones in the oceans.

The large bio-molecules that make up a cell are composed of long chains (polypeptides) of simple units made mostly of carbon and hydrogen atoms. The billions of different proteins in all living things are chains of hundreds or thousands of just twenty different kinds of amino acids. These amino acids, fairly simple molecules composed of a few tens of atoms, are the same in all life forms. The important nucleic acids in all living things are all built of chains of just five different nucleotides, which also are fairly simple mol-ecules. Polysaccharide molecules, such as starch and cellulose, are chains of simple sugars, like glucose. Lipids, the components of cellular membranes (also used to store energy), are also long chains of basic simpler units, and are the same for all organisms.

The complex biochemical transformations taking place in the cells of living matter are accelerated by enzymes, specialized proteins needed to build and maintain a cell. These enzymes catalyze a great variety of biochemical reactions in the cell, promoting the rearrangement of simple compounds into the complex bio-molecules of life, or mediating the destruction of complex molecules once they are used. Enzymes control cell growth, determine cell function, control physiological processes and play crucial roles in the development of organisms. As catalysts they promote reactions, increasing their rates by factors of millions without themselves being altered in the process. Without enzymes the reactions would be so slow that life would not prosper.

Living systems, even the simplest of organisms, must achieve the high level of biochemical sophistication that we know as life to be alive. Even a simple organism such as the (usually) friendly bacterium *Escherichia coli*, the most thoroughly studied organism in the world, is built of about 5000 different organic compounds, including 3000 different proteins and 1000 different kinds of nucleic acids. So simplicity is only relative; the bacterial cell is fantastically complicated. The 10 000 billion eukaryotic cells in our bodies are each 10 000 times larger than *E. coli*, with as many as 5 million different kinds of proteins, themselves different from those in *E. coli*, all made with the same twenty amino acids. But how is this all organized? Who worked on the blueprints to produce these very complex systems?

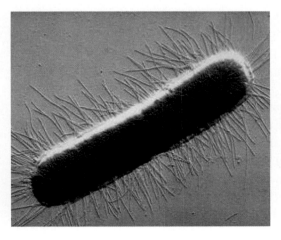

The bacterium *Escherichia coli* lives in symbiosis with us. It helps our digestive process, and produces vitamin K and B-complex in our intestines. It was discovered in 1885 by the German bacteriologist Theodor Escherich (1857–1911). However, a mutated strain of *E coli*, called O157:H7, incorporates a gene which makes the bacterium produce a powerful toxin – Vero toxin. This can lead to severe illness producing stomach pains and bloody diarrhea, and in some cases death. Cook your hamburgers well-done, don't drink untreated water, and be very clean in your kitchen. (Image courtesy of Indigo Instruments (www.indigo.com))

The secret of life

The famous double helix of DNA is the macromolecule that contains the secret of life. We wrested part of this secret when we discovered that the blueprint containing all the information needed to build an organism is encoded in its *genome* and stored in the DNA of every cell. We also discovered how this information is replicated when a cell divides, and how the information is encoded and used to build an organism. How this came to be, how this information was initially encoded, is one of the deepest mysteries about life, a complex problem which will only slowly yield its secret to enquiring minds.

All living things, be they a tiny bacterium or a whale, contain the same type of nucleic acids: DNA (*deoxyribonucleic acid*) and RNA (*ribonucleic acid*). The large DNA molecule is the repository of the genome of a species, containing the total genetic information needed to construct, from one cell, another cell or the entire organism. Two strands, each formed by a long chain of four distinct molecules called nucleotides, form the DNA molecule. Each nucleotide contains one of four bases: adenine (A), guanine (G), cytosine (C), and thymine (T). The double helix is built as nucleotide bases form weak chemical bonds – produced by electrical forces between atoms – in pairs: adenine with thymine and guanine with cytosine.

When a cell divides, its DNA molecules replicate. The replication proceeds by unwinding the two long strands which make up the double helix, and separating the base pairs that hold them together. Each strand now serves as a template to build a new strand, using available nucleotides and aided by the action of enzymes. This results in two identical double helices as shown in the figure on page 136. This process of replication is the *fundamental mechanism of life*. The RNA molecule is similar to DNA but smaller and single stranded, and the base uracil (U) substitutes thymine. RNA forms the bulk of the nucleic acids found in all cells, carrying out many functions related to the production of proteins, the basic building blocks of all organisms. Only in viruses is the genome sometimes stored in RNA instead of the more complex DNA.

All this sounds quite complex, and it is, when looked at in detail. However, what is essential is not so complicated. These giant molecules are simply the way to conserve and duplicate, generation after generation, the information needed to build organisms, molecule by molecule, cell by cell. Hereditary information is encoded along the DNA strand in the sequence of nucleotides which make up the alphabet of the genetic code. The sequence of millions of A, G, C, and T along the DNA molecule are read three at a time (a codon), specifying one amino acid of the sequence that forms a protein. Thus, for example, the codon CTA specifies the amino acid leucin. This was decoded in the 1960s, a momentous achievement. A gene is the basic unit of inherited information, a segment of the DNA double helix that specifies a particular trait through the manufacture of a selected set of proteins, or serves as a regulatory agent. Genes can specify the color of the eye, or the structure of a particular protein, or can turn on, or off, certain steps in a biochemical process taking place in the cell. As cells divide, genes will determine how new cells will differ from their parent cells, leading to the development of the different tissues which compose a complex organism. The genes of a tadpole will at the right time start processes to transform it to a frog. Our genes will at the right time change the voice of a boy into that of a man and, in a woman, start the process of menstruation which happens on average every 28 days, the past synodic period of the Moon, as we have seen.

Several hundred macromolecules – RNA and enzymes – are involved in the production of proteins in a cell, a process that consumes about 90 percent of the energy produced in a cell. A protein made up of about 100 amino acids can be synthesized in about five seconds. At any time, a cell is making thousands of different proteins. For example, every second, the cells in your bone marrow produce 1000 billion hemoglobin molecules, and you do not feel any of this. Hemoglobin is a protein in your red blood cells, built of 574 amino acids, which carries oxygen from your lungs to the rest of your body and waste CO_2 from the cells back to your lungs to be exhaled. One drop of blood contains 500 million red blood cells.

The DNA molecule in a virus is made up of a few thousand nucleotides. Simple bacteria such as *E. coli*, and one of our greatest enemies, *Mycobacterium tuberculosis*, contain more than 4 million base pairs. It only recently became possible to determine their complete nucleotide sequences, opening the way to identify every major gene function. *M. tuberculosis* is also known as Koch's bacillus in honor of its discoverer (in 1882) and founder of modern bacteriology, Robert Heinrich Hermann Koch (1843–1910), who received the 1905 Nobel Prize in Medicine for this discovery. The knowledge of the genome of a pathogen (a bacterium that can cause disease) will give us an opportunity to find ways to defend us from it. The fruit fly, *Drosophila melanogaster*, has been studied for many years to understand the genetics of more complex multicellular organisms. Its genome, 50 times more complex than that of a bacterium, composed of about 180 million base pairs, has recently been completely sequenced.

The Human Genome Project is a huge endeavor that will take several

THE DNA GENETIC CODE

		Second base in codon			
		T	C	A	G
First base in codon	T	TTT Phe TTC Phe TTA Leu TTG Leu	TCT Ser TCC Ser TCA Ser TCG Ser	TAT Tyr TAC Tyr TAA *Stop* TAG *Stop*	TGT Cys TGC Cys TGA *Stop* TGG Trp
	C	CTT Leu CTC Leu CTA Leu CTG Leu	CCT Pro CCC Pro CCA Pro CCG Pro	CAT His CAC His CAA Gln CAG Gln	CGT Arg CGC Arg CGA Arg CGG Arg
	A	ATT Ile ATC Ile ATA Ile ATG Met, *Start*	ACT Thr ACC Thr ACA Thr ACG Thr	AAT Asn AAC Asn AAA Lys AAG Lys	AGT Ser AGC Ser AGA Arg AGG Arg
	G	GTT Val GTC Val GTA Val GTG Val	GCT Ala GCC Ala GCA Ala GCG Ala	GAT Asp GAC Asp GAA Glu GAG Glu	GGT Gly GGC Gly GGA Gly GGG Gly

Bases: A = adenine G = guanine C = cytosine T = thymine
The 20 amino acids specified by each codon

Ala: Alanine	Leu: Leucine
Asp: Aspartic acid	Lys: Lysine
Arg: Arginine	Met: Methionine
Asn Asparagine	Pro: Proline
Cys: Cysteine	Phe: Phenylalanine
Glu: Glutamic acid	Ser: Serine
Gln: Glutamine	Thr: Threonine
Gly: Glycine	Trp: Tryptophane
His: Histidine	Tyr: Tyrosine
Ile: Isoleucine	Val: Valine

Notes

The most common amino acids such as leucine are coded by several different codons while the rare tryptophane is coded only by one.

Codes for chemically similar amino acids are similar, as is the case for aspartic acid and glutamic acid.

Similar codons specify the same amino acid, and for some the third letter is redundant as for glycine.

years of work. Its goal is to obtain the entire sequence of nucleotides in the human genome, approximately 3 billion base pairs, and understand the function of all the estimated 40 000 genes distributed among the 23 pairs of human chromosomes. On November 17, 1999, the one-billionth base pair of human DNA was reached – it was a "GC." This international enterprise will bring us closer to understanding and finding cures for the many diseases which affect us, and by the time you read this the sequencing might be complete, although its interpretation will take much longer. At a more fundamental level it will discover our genetic heritage and what makes us different from other animals, in other words, what makes us human.

If all the DNA in a human cell were laid out along a straight line, the molecule would be about 6 feet long, but it is extremely thin. If it were a thin wire it would stretch to about 50 miles. You could write the code on one million pages containing just As, Cs, Gs and Ts. You would, of course, not do this but instead store the entire thing on a few CD-ROMs. Human DNA is a very large molecule yet it fits into the small nucleus of a cell, much smaller than the dot on this *i*, only because it is so thin and can be folded and coiled about the proteins in a chromosome. That all information needed to build a complex organism can be stored within this tiny dot is truly amazing, a discovery made possible by the scientific and technical developments of recent years, which have given us the power to explore the farthest reaches of the Universe and the most intimate submicroscopic structures within the Universe of a cell.

We are understandably impressed by our state of knowledge. Perhaps 1000 years from now someone (I can't imagine who) will look at us in the way that we look at those from 1000 years ago. They will wonder how we did not know something as fundamental as . . . I wish I could fill in the blanks. But, for the moment, these are fantastic discoveries.

The 1962 Nobel Prize in Physiology or Medicine, was awarded to James Watson (1928–), Francis Crick (1916–) and Maurice Wilkins (1916–) for working out the structure of the DNA molecule. This is the basis for understanding its replication, the process of protein synthesis, and the transfer of genetic information, the greatest achievement of the twentieth century's work in the life sciences. By 1966 the genetic code was finally cracked as a result of the work of two independent groups, one led by Har Gobind Khorana (1922–) and the other by Marshall Warren Nirenberg (1929–). For this they were awarded the 1968 Nobel Prize in Physiology or Medicine, which they shared with Robert Holley (1922–1993) who determined the mechanism by which the code is translated into an amino acid sequence by means of transfer RNA.

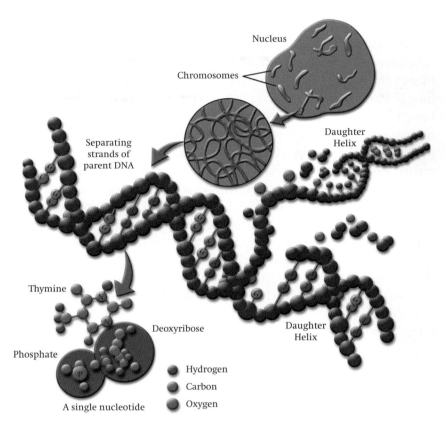

The fundamental process of life
(a) Each cell of the human body contains 23 pairs of chromosomes in the nucleus (except the germ cells which do not contain pairs), each a packet of compressed and entwined DNA and protein. Every DNA strand is a huge natural polymer of repeating nucleotide units, each of which comprises a phosphate group, a sugar (deoxyribose), and a base (either adenine, thymine, cytosine, or guanine). Every strand thus embodies a code of four characters (As, Ts, Cs, and Gs), the blueprint for the machinery of human life. In its normal state, DNA takes the form of a highly regular double-stranded helix, the strands of which are linked by hydrogen bonds in pairs between adenine and thymine and between cytosine and guanine. The human genome is composed of some 3 billion base pairs. It is the specificity of these base-pair linkages that underlies the mechanism of DNA replication illustrated here. Each strand of the double helix serves as a template for the synthesis of a new strand. Replication thus produces twin daughter helices, each an exact replica of its sole parent.

From bacteria to humans, we all use the same code (although a few exceptions have been discovered), this having remained unchanged for billions of years and over uncountable generations. The universality of the genetic code, which could have been different for different organisms, and the universality of the process of protein synthesis, are proof of its descent from one

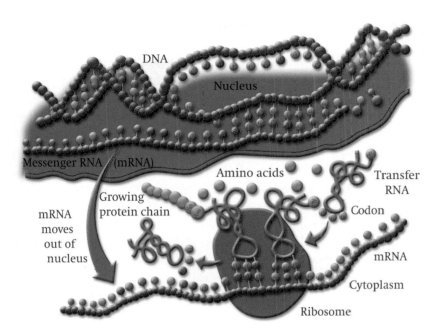

(b) In the nucleus of a cell, RNA is produced by transcription, in much the same way that DNA replicates itself. RNA, however, substitutes the sugar ribose for deoxyribose and the base uracil for thymine, and is usually single-stranded. One form of RNA, messenger RNA (mRNA), conveys the DNA recipe for protein synthesis to the cell cytoplasm. There, bound temporarily to a cytoplasmic particle known as a ribosome, each three-base codon of the mRNA links to a specific form of transfer RNA (tRNA) containing the complementary three-base sequence. This tRNA, in turn, transfers a single amino acid to a growing protein chain. Each codon thus unambiguously directs the addition of one amino acid to the protein. On the other hand, the same amino acid can be added by different codons; in this illustration, the mRNA sequences GCA and GCC are both specifying the addition of the amino acid alanine (Ala). (José F. Salgado after *To Know Ourselves*, 1996, Human Genome Program, US Department of Energy, E. O. Lawrence Berkeley National Laboratory)

origin. The underlying uniformity observed in all living systems points to *the fundamental oneness of life* and indicates that wherever life arose on Earth it was one and the same process. We shall know if the code is really universal if we find life elsewhere, although if life is found elsewhere in the solar system we would first have to rule out a common origin by some way of "contamination" between the planets. We would be certain if someone from very far away, meaning another planetary system, tells us what their code is. The universality of the code would be a momentous discovery, since it would prove that the origin of life follows some as yet unknown laws, and is not a random process.

Evolution

The science of molecular biology has revolutionized our understanding of life and has instilled order into the bewildering biological diversity we observe. Without doubt, it is the science of the new century. It all began when an Austrian Augustinian monk, Gregor Johann Mendel (1822–1884), discovered the laws of inheritance which are at the foundation of genetics. These laws are fundamental to understanding the process of life's evolution as developed by Charles Robert Darwin (1809–1882) and Alfred Russell Wallace (1823–1913). Mendel obtained his results from experiments with peas planted in the garden of his monastery, in the small town of Brünn (now in the Czech republic). He published these in 1866 in the transactions of the Brünn Natural Science Society. There they remained, almost unknown to the rest of the world, until discovered 34 years later, in 1900. Neither Wallace nor Darwin knew of Mendel's results.

Darwin was born on February 12, 1809 in Shrewsbury, England. That same day Abraham Lincoln was born in Hodgenville, Kentucky. In December of 1831, at age 22, some 300 years after Magellan, Darwin embarked on the frigate *HMS Beagle* for his famous five-year expedition to explore the southern parts of South America and its Pacific coast, including some islands of the Pacific, notably the Galapagos. After crossing the Pacific ocean, rounding Africa and stopping again in Brazil, he finally returned home in October of 1836. His observations during this trip provided the material to write his most famous book, published in 1859: *The Origin of Species by Natural Selection or the Preservation of Favoured Races in the Struggle for Life*. In 1871 his *The Descent of Man and Selection in Relation to Sex* was published.

Wallace was born in Usk, Monmouthshire, on January 8, 1823. He traveled widely, first to the Amazon and later to the Malay Archipelago. In 1855 he published the essay: *On the Law which has Regulated the Introduction of new Species*, arriving independently at the same conclusions as Darwin. According to Darwin and Wallace, men (and women) were not specially created by the grace of God, but descended from the animal world through a long and complex chain of events.

The theory of evolution, which explains the observed facts, had (and has) as much trouble being accepted as that of Copernicus and for similar reasons. However, resistance to it was intensified because of a more personal affront to our collective self-esteem. We were not at the center of the Universe, we were not even at the center of creation and, for the majority of history, life was not even concerned with us. Evolutionary ideas were contrary to a static view of our history and of society as represented by the ideas of the influential Swedish naturalist Carolus Linnaeus (1707–1778). He set out to classify all of life assuming that every individual in a given species could be traced back to an original pair produced at the creation, quite a challenging task. Such ideas are no longer considered valid, but his system of nomenclature is still that in use today. The manuscripts of Linnaeus and his collections were bought by James Smith (1759–1828) in 1784 and are carefully preserved at the Linnean Society of London, which Smith helped found in 1788. It was here that Charles Lyell, helping to avoid a sorry dispute over precedence, arranged for Wallace and Darwin to present their work jointly on July 1, 1858. Symbolically, this date marks an important turning point in the history of thought about life on Earth.

Darwin died on April 19, 1882, and was buried in Westminster Abbey, joining, among other famous thinkers, Newton. Wallace died at Broadstone, Dorset, on November 7, 1913, where his remains are kept in a neglected grave at the cemetery. Such is life.

Even today, as we start this new millennium, there is fierce opposition by some to the theory of evolution, which is seen as contrary to a literal interpretation of biblical genesis. There have been places where school boards have attempted, sometimes successfully, to ban evolution from the curriculum – the Inquisition in a modern, less cruel form. They might as well legislate that the stars are eternal, that 2 times 2 is 5, and that the Earth does (again) not move. However, the theory of evolution has nothing to say about biblical scriptures, any more than Newton's theory of gravitation does. Nor do the scriptures tell us anything about our physical world; in the words of Caesar Cardinal Baronius (1538–1607), historian of the Catholic Church: "The Holy Ghost intended to teach how to go to heaven, not how the heavens go."[2]

We distinguish between the fact that evolution occurs and the theories as to how evolution proceeds, just as we distinguish between the fact that the planets orbit the Sun in elliptical orbits and the theory of Newtonian gravitation. Someday the theory of evolution might be modified as new facts are uncovered, but this will not mean that it is wrong, just as Newton's theory is not wrong in spite of some new facts which led to its revision by Einstein. Apples will continue to fall toward the center of the Earth and fossils will continue to appear to show us in ever-increasing detail the history of life upon our planet.

The theory of evolution does allow us to understand important aspects

[2] Quoted in Jerome Langsford, *Galileo, Science and the Church*, St Augustine Press, Indiana, 1998.

of the biological world we observe and are part of, and in particular our origin, without the need to invoke supernatural forces. It has unified the biological sciences in the same way that Newton's theory unified the physical sciences. Although scientists argue about such details as the rate of evolution, the importance of different mechanisms which generate new species (speciation), and the precise genealogy of all organisms (phylogeny), the essence of the theory is surprisingly simple: the variations between individuals of a species, reflecting genetic differences, can accumulate over many generations. After a population is reproductively isolated, say by geographical barriers produced by plate tectonics, such as new bodies of water or mountain ranges, the accumulated variations will produce differences such that the two separated populations can no longer interbreed – the process of speciation. The most important mechanism which determines the final outcome is natural selection, which will favor those variations which provide an individual with an advantage to survive and multiply within the existing environment and in this way propagate its genes – *survival of the fittest* – producing adaptation. The variations are due to an accumulation of changes in the genetic information which are inherited following the rules of genetics.

> *Survival of the fittest*
> Two explorers were walking along a narrow path inside a dense jungle when, after a turn, they found themselves face to face with a ferocious lion. Without thinking twice, the first explorer placed his heavy equipment on the ground. "What are you doing?" asked the other. "I am going to run" responded the first one. "Don't be silly, nobody can outrun a lion" said the other. "It isn't necessary" said the first one; "I only need to run faster than you," and started running.

Today our species inhabits the entire planet and it is no longer possible to find space to isolate a population reproductively. Our evolution by natural selection has come to an end, although we continue to evolve culturally. Evolution by natural selection is the mechanism which produced the great variety of organisms we observe, a process which makes it possible for populations to adapt over time to the always-changing environment on Earth, and to evolve new forms as circumstances and luck will have it.

Yes, *luck*. There is no sense of purpose in evolution, no long-term goals, no design, just a series of experiments driven by random changes in the nucleotide sequences in the DNA of the gene of some organism and therefore introducing random changes in the outcome. The environment determines who will be successful. Plant and animal breeders use artificial selection to obtain new varieties, they choose what they want from a random collection of results. Similarly, nature chooses what is going to succeed. The important difference is that breeders have a plan, a goal to achieve, whereas natural selection is blind and does not care about the result.

Changes in the nucleotide sequence – mutations – can be caused by many agents such as ultraviolet radiation, X-rays, radioactivity, or chemical agents, and even cosmic rays arriving from distant supernovae to affect the evolution of life. Thus, in this indirect and somewhat poetic way, the stars that created the elements of life and provide the energy for life, also influence its course. Sometimes mutations are simply a consequence of errors in the DNA copying process. Changes in the DNA of a gene will result in changes of the resultant protein, or changes in some regulatory function with usually lethal consequences for the organism. Only a very small number of these changes result by chance in offspring having a characteristic that makes them better at surviving and reproducing in the environment they encounter, including competition with others for the nourishment necessary to survive. *Without mutations there would be no evolution.*

Darwin and Wallace completed the, for some painful, displacement of humans from a central position and role in the Universe to nothing more than the result of some accidents, and much opposition is simply due to this hard-to-swallow fact. Still,

> There is grandeur in this view of life, with its several powers, having been originally breathed into a few forms or into one; and that, whilst this planet has gone cycling on according to the fixed law of gravity, from so simple a beginning endless forms most beautiful and most wonderful have been, and are being evolved.[3]

So ends Darwin's *Origin of Species*.

Sometimes you might read that evolution is "just a theory," as if any other theory would do, and as if it were just a matter of taste. Well, it is not. A scientific theory is supported by experimental evidence, by facts we observe, and will be modified or rejected if the facts are counter to it. A common usage of the word "theory" means "guess," quite different from its meaning in science. Newton's theory is not "just a theory." It correctly describes what we observe on Earth and in the Heavens, and provides a framework for understanding the physical world. Nobody now disputes that

---

[3] Charles Darwin, *On the Origin of Species by Means of Natural Selection, or the Preservation of Favoured Races in the Struggle for Life*, John Murray, London, 1859.

planets move about the Sun in elliptical orbits, and nobody in their right mind who has studied the facts can dispute that evolution does occur. We use the laws of Newton to predict a spacecraft's landing spot on the Moon, and by Newton it gets there. You will also find a lot written, even by "respectable" members of our society, using some part of evolutionary science to prove the superiority of their kin. Not surprisingly, they never arrive at the opposite conclusion.

Evolution predicts that mutations will make it difficult to conquer infectious diseases and fight the pests which destroy our crops. Antibiotics and vaccines work against specific agents and, as these are eliminated, some mutated forms which show resistance become fit to survive and multiply in the environment created by these antibiotics or vaccines. This is "survival of the fittest" which, because of the inheritance of these new traits, will evolve into a new strain of bacterium or virus. This we are acutely aware of, as every year about 50 million of us catch the flu. There is nothing theoretical about that.

Apart from getting us into all sorts of trouble, the invention of sex changed the rate of evolution. Before this, reproduction was simply achieved by cell division, as it still is for bacteria and for the cells of our bodies, and the only differences between parents and offspring were those induced by the occasional viable mutation. If you've seen one *E. coli*, you have seen most of them. Sexual reproduction, by combining the different genomes of the parents, mixes genetic information and produces offspring that differ from both parents, with different genes which again produce offspring which are different, and so on, leading to variations and the possibility of rapid evolution. During meiosis – the special cell division which produces the sperm and egg – chromosomes can exchange genes and form new combinations so that the fertilized eggs (the zygotes) are never exactly the same (except for identical twins). This is why we are all different, all 6 billion of us. If at some point a small population becomes reproductively isolated, it can diverge until eventually it becomes a new species. If you see one person, you have only seen that one and, when that person dies, a unique individual unfortunately disappears forever.

Origin

So, how on Earth did life originate? The particular way that life arose, in other words, going from the raw materials of life, possibly delivered by cometary bombardment, to the production of a cell, or even something more primitive like a virus, is a fundamental question. Whatever happened, it must have happened in less than 300 million years, a relatively short time interval. In 1871, in a letter to a friend, Darwin wrote: "if (and oh! what a big if!) we could conceive in some warm little pond, with all sorts of ammonia and phos-phoric salts, light, heat, electricity, etc, present, that a protein compound was chemically formed, ready to undergo still more complex changes . . ."[4]

We have learned a lot in the 130 years since Darwin wrote this, but in regard to the origin of life our level of understanding is not much better than his. The question about the transition from inanimate matter to living things is a profound and difficult one, the answer to which we would dearly like to know. An examination of living things shows us that they are all built from ordinary molecules that conform to all the known laws of physics and chemistry. However, when these materials are put together to form a living organism, properties appear that are obviously not part of inanimate matter. The whole seems larger than the sum of its parts, a property that still needs a satisfactory explanation. The question about the nature of our con-sciousness and intelligence is particularly difficult to answer. We have a system of great complexity such as the brain, and this system can think in abstract terms and be aware of itself and its place in relation to everything else. Our individual thoughts seem to transcend our physical existence, and acquire independence. They can remain after we are gone, recorded in per-manent records such as books, to become the thoughts of our species. It is not clear if consciousness is understandable in terms of the physical laws the brain has so far discovered, or will discover in the future. In other words, the question is: can the brain understand itself? The answer might be that it cannot, a mystery forever hidden from us.

Damaging ultraviolet radiation will not penetrate water beyond a few meters, so life would be safe from this radiation, even at the earliest time when the atmosphere was devoid of ozone. Therefore it is logical to think that life developed first under water, and only later migrated to land, as indeed the fossil record indicates. We have already seen that liquid water is essential for life; we cannot survive for long without it, and it would have been even more important at life's origin. Maybe the extremophiles are giving us a hint: life arose somewhere deep in the oceans of the archaean. There it would have been relatively protected from the violent events at the surface, and able to survive the harsh conditions of the primitive environ-ment. Indeed, genetic studies indicate that the ancestors of modern bacte-ria were extremophiles.

---

[4] Quoted in Robert M. Hutchins (ed.), *Great Books of the Western World, Vol. 21 Darwin*, Encyclopaedia Britannica.

We think we know roughly when and generally where life began, and we are looking to see if it also might have happened elsewhere, but *how* it began still eludes us. Understanding how life began is one of the most exciting challenges we face. Of course, one can say, "Then a miracle happened and . . ." but we do not like miracles as place-holders for our ignorance. It is better to say that we do not exactly know, a good reason to keep on researching with the confidence that in the same way that several other past miracles were finally understood, this one will also eventually yield to the inquiring minds of scientists. Although we understand the complex biochemical processes used by life and also understand the organization and evolution of living systems, we do not know how life originated. It is fair to say that, until we know this, we cannot really claim with certainty that we know what life is, and our attempts at defining it will of necessity remain incomplete.

In 1953 Harold Urey (1893–1981), a physicist who had been awarded the 1934 Nobel Prize in Chemistry for his discovery of deuterium, and Stanley Miller, then his graduate student, showed that it was possible to synthesize amino acids from a mixture of methane, ammonia, and water when it was subjected to electrical discharges. Since this pioneering experiment, many others have shown that under conditions simulating the early environment of Earth, a large variety of small organic molecules of biological significance can be created. But some of the larger, more complex molecules are not as easy to produce. It does seem that the elaborate series of steps required to synthesize some of the more complex bio-molecules which constitute living systems, from inorganic material, are impossible to achieve naturally by a *random process*. It is harder still to produce a complex system of these molecules which will organize spontaneously into a self-replicating entity. There seems not to have been sufficient time and space for such an apparently improbable event to happen by chance.

This has led to the idea that an initially simpler system, an ultimate ancestor to archaea and bacteria, traces of which we have not yet found, and might never find, was produced by chemical synthesis and then evolved to the more complex ones we observe. Even then we are led to the uncomfortable conclusion that life is an extremely improbable event, something which would not happen again in a billion years on a billion Earths. Of course, we are then back to being unique and very special, which might make us feel good in an odd way, but which makes the Universe and all that happens there, all the stars and all the supernovae in the millions of galaxies, a strange and lonely place, "mostly harmless." To me this is a frightful perspective.

Most likely there is something we have not yet discovered that will show that the origin of life is not the result of random events, and that it follows some natural rules which make it not so unlikely, therefore showing how easily life arises under the right conditions. This will leave me with a much better feeling. Over the history of science, events which at one time seemed inexplicable were later seen to be the consequence of some natural law; we don't need angels to move the planets. This is what keeps scientists interested in doing science.

We shall be closer to an answer if we find some evidence of present, or even past, life elsewhere in our solar system. I do not mean intelligent life – that is a different ball game – I mean just life: a bacterium will do. This would provide us with the unique opportunity to study these different life forms comparatively, allowing us to discover some of the general underlying principles which lead to life. Because of the importance of water for life, the search for life on other worlds is equivalent to a search for water or evidence of past water.

The long and winding road

Life on Earth remained simple and microscopic for a very long time, spanning about the first 3 billion years of life's history. While it took less than 300 million years to pass from inanimate matter to life, it took an exceedingly long time for life to go from microscopic unicellular organisms to multicellular ones. Why this was so we do not know. It was an important transition that opened the way for an entirely new form of life, one based on cooperation.

The first aerobic organisms appeared about 2 billion years ago, when the geological record shows evidence of oxidation on land, a consequence of sufficient oxygen accumulation in the atmosphere. The more complex eukaryotic single-celled organisms, the microscopic protists, developed toward the end of the Mesoproterozoic, about 1 billion years ago. It took several hundreds of millions of years for life to go from the protists to the first small multicellular animals, the metazoans, but the exact sequence is lost in the fossil record. The oldest eukaryotic multicellular organisms – plants, animals, and fungi – appear in the fossil record about 550 million years ago.

Then, about 540 million years ago, an event known as the Cambrian explosion occurred, which marks the start of the greatest diversification in the history of life that we know of and the first appearance of animals large enough to be seen without microscopes. Although it took about 10 million years, this was an "explosion" geologically speaking. It was the "Big Bang" of biology. It is quite likely that this development was connected with the increased availability of free oxygen in the atmosphere, which by that time had reached modern levels. It is also likely that the great burst of diversification was related to the fact that the ecological landscape was empty, and any new anatomical designs had a good chance to develop, without a great need to compete with others or worry about predation. All phyla of modern life, that is, categories of organisms that have unique anatomical designs, make their first appearance in the fossil record at this time. The first (now extinct) trilobites appear here and also the brachiopods with shells like modern clams. We, the mammals, belong to the phylum of the chordates, which also make their first appearance in this period, and includes the vertebrates. These are characterized by the spinal cord, the nerve trunk that distributes signals between the brain and the body.

The Cambrian fossils of the Burgess Shale, a limestone quarry high in the Canadian Rockies of British Columbia, were among the most exciting finds in the history of paleontology, as they preserve a precious record of soft-bodied marine animals, a rare event in the fossil record. These fossils were preserved because they were rapidly buried under mud in an oxygen-free environment, so that they did not decay. They were discovered by Charles Doolittle Walcott (1850–1927) in 1909, then secretary of the Smithsonian Institution in Washington, DC. He is an important figure in the history of paleontology, and was a very successful science administrator, becoming president of the National Academy of Sciences and a close advisor to presidents from Theodore Roosevelt to Calvin Coolidge. Walcott is also credited with establishing the fact that a significant Precambrian fossil record did exist, demonstrating the astonishingly long history of life. Darwin had wondered, and worried, about Precambrian life and was concerned about the lack of any fossils from that time. We now know that Precambrian life was microscopic, so it is no wonder that its remains were not found until Walcott's pioneering work.

The Burgess Shale fossils preserve for our study the remains of a large and unique variety of animals that lived 540 million years ago. This includes the remains of animals such as *Opabinia*, a five-eyed animal a few inches in length with no modern descendants, and *Hallucigenia*, a bizarre animal 1 inch in length, which looks as if it belongs in a science fiction movie, and which also seems to have no modern descendants. There is also *Pikaia*, the world's first known chordate, which at least symbolically stands for our oldest ancestor, the first of the phylum to which all the vertebrates, fish, reptiles, amphibians, birds, and mammals, including ourselves, belong. The Cambrian period is followed by the Ordovician, the end of which, 440

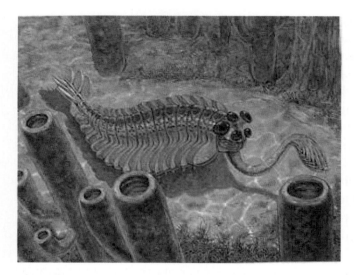

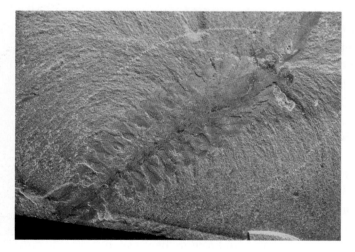

The Burgess Shale fauna discovered in 1909 in the western Canadian Rocky Mountains, preserves for us the memory of life on an alien planet: Earth as it was 500 million years ago, early in the Cambrian period. In those days the site was located in warm tropical coastal waters near the equator. Plate tectonics lifted and moved the site to where it is now at a height of 8000 feet. The fossil of this strange *Opabinia regalis* was left by a small – 3 inch long – animal with five eyes and a miniature trunk ending in a claw-like structure used to grab things. (Art by David Miller, fossil © the Peabody Museum of Natural History, Yale University, New Haven, CT)

million years ago, is marked by a mass extinction. Had *Pikaia* or its descendants not made it through this transition, this story would have turned out quite differently.

By 450 million years ago, at the Ordovician, we find the first land plants, descendants of multicellular algae. Before this, the land had been a barren inhospitable place. Ozone now made life on land possible, and after some 50 million years the stage was set for an important transition. Animals, until then living under the protection of water, ventured perilously onto land. Perhaps there still is a memory of this epic moment stored in our cells. Perhaps this is the reason why, as you feel the warmth of the Sun when you are lying on a beach, the sound of the surf, even its smell, soothes your spirit.

As organisms evolved from some common ancestor, the differences in the nucleotide sequences in the corresponding genes of each descendant species increased with time, so that the history of life's evolution is written in the genes. Therefore, a comparison of the nucleotide sequences in the DNA of corresponding genes of different species provides a measure of kinship: the smaller the difference, the closer the relationship. These studies complement those of paleontology and anatomy and leave little doubt about the descent of animals (including us) and plants from other simpler life forms, although many details remain to be filled in. Results of these types of studies in molecular evolution have shown that over 98 percent of the genome of humans and chimpanzees have similar sequences. It is surprising that the large difference between us and them is the result of only a 2 percent genetic difference. These apes *are* our cousins. So next time you go to the zoo sit and look them in the eye. A haunting feeling will propagate through your neurons.

Sixty-five million years ago, a terrible cosmic accident caused a mass extinction that ended the age of the dinosaurs, as we shall see in the next chapter. This accident gave the insignificant mammals the opportunity to develop into the rulers of the vertebrate world, finally to become the dominating species on Earth: *Homo sapiens*. Some 5 million years ago, the motion of the continents that produced the Great African Rift Valley changed the climate in Africa. Humid tropical forest changed to savannah, and a primate with the ability to walk upright – *Australopithecus*, "southern ape" – was able to run faster and see farther, an advantage to survive in the savannah. The famous 3 million year old "Lucy" is one of the best preserved remains of *Australopithecus afarensis*, found in Ethiopia in 1974. Walking upright was a vital adaptation, freeing the hands and paving the way to put them to good use. Less than 2 million years ago, some think much less, there occurred what is possibly the most significant evolutionary transition of all times. Descendants of those hominids who had left their footprints in Laetoli developed large brains, developed the ability to make tools, and mastered the use of language. This changed the character of life forever. Whereas, before this, life had been a mindless endeavor, going about the business of living without much or any thought, it was transformed into a self-conscious phenomenon which wants and needs to know. *The world was given a mind*. This has made all the difference, despite the fact that we collectively act in a mindless way.

We know little about the origin and nature of consciousness and intelli-

Our closest relative: the chimpanzee (*Pan troglodytes*). (© SteveBloom.com)

gence. It seems reasonable to assume that the first simple organisms did not have these qualities, yet we do. The question then is where, among the different forms of life, is the line crossed between unconscious and conscious life, between non-intelligent and intelligent life, and what determines this? We have no clue, although it is clearly related to the complexity and size of the brain. The scant fossil record shows this as we go from *Australophitecus* (brain size about 50 cubic centimeters), passing through *Homo habilis* (750 cm^3) and *Homo erectus* (1000 cm^3), our direct ancestors, to *Homo sapiens* (brain size 1400 cm^3). Overall, the brain increased in size by a factor of 3.

Contrary to what many would like, we are not the result of some design, of some necessary move toward consciousness and intelligence. We are the result of an amazing process that just happened, though the world would have been just as happy with only bacteria and, sad to say, maybe happier. Our blueprint, the result of trial and error, was designed by the firm *Mother Nature – Architects and Engineers*. This might seem depressing, since what I am saying is that our existence is just a fluke, the result of an intricate web of events, which would never develop again in the same way if we started our world anew 4 billion years ago. *Pikaia* might not have made it, or maybe the dinosaurs would still be around, or a myriad other things could have evolved differently. It remains to be seen if we make it, although I can't imagine by whom it will be seen.

If you get the feeling that my message here is that we are not important in the grand scheme of things, then you *are* getting the message. Well, our self-esteem has been badly shaken before when we learned that the Earth was not at the center of the Universe, or even the solar system, or even our Galaxy, which is not even ours. Our planet is an insignificant microscopic speck of dust in the cosmic landscape. For all we know, tomorrow we might be wiped out by a virus, or by an asteroid crossing our path around the Sun or, what is worse, by our own doings. What would happen after that?

The Universe is so large that, even if it is improbable on any particular planet, sentient intelligent beings might be somewhere, assuming that life arises with some ease, even if rare and hard to find. These beings may be so intelligent that if we met them we would feel like the apes we really are. If we feel insulted by this monkey business it is simply because of our myopic view of things when we presume that we are the crowning achievement of something, really the best, with a subjective definition of "best" that is in great measure the reason for our arrogance. If you do not believe me then ask a monkey for her opinion.

No one can deny that our brain power is supreme when compared with any other animal on Earth, and in this sense we are unique and privileged. However, this does not make us any better than the rest, nor does it give us the right to eliminate them from the face of the Earth. When we look at what we have done with this power, we can be proud of many achievements, material and intellectual. However, at the same time we have shown that we are capable of the most evil acts of horrifying cruelty, not seen anywhere else in the animal kingdom and, we hope, not to be seen anywhere else, ever. There is nothing to be proud of there. It is time to get off this great collective ego-trip and look at ourselves from a different perspective.

Humans have tried to assign meaning to life by asserting in one way or another that all this was meant for us, either because some supernatural will decreed it, or at least because nature is designed in such a way that it will end with us. *If we are not careful, it really will end with us.*

Ah, yes, I meant to tell you that after 15 hours the total number of bacteria in the example on page 115 is 1 073 741 824, which is 2 to the power of 30.

(NASA, art by Don Davis)

Chapter 6

Close encounters of many kinds

Relax, don't panic[1]

Dangerous neighborhood

It takes only a good long look at the night sky to know that Earth is constantly being bombarded by small objects which fortunately burn up as they enter the atmosphere at high velocity. Larger projectiles are also in space, lurking in the dark, fewer in number but extremely dangerous. We will clearly be hit again, as we have been in the past, but we do not know when or what the effects will be.

Our Sun contains 99.8 percent of all the matter in the solar system, in other words, almost all of it. As we saw in Chapter 3, the rest – mere leftovers from the material which condensed out of the disk that formed the Sun – produced the planets, their satellites, the comets, and the asteroids. Of this leftover material, Jupiter contains more than twice the matter in all the other planets combined. Indeed, the Earth is only the leftover of the leftover, a small planet which happened to form at the right place for us.

Except for Pluto (of dubious credentials), all planets orbit in nearly the same plane and in the same direction, a natural consequence of the process of formation of the solar system. This plane coincides with the equator of the Sun's rotation, which is also in the same direction, once in about 26 days. However, being a gaseous object, the Sun rotates faster at the equator than at its poles. Most asteroids follow elliptical orbits around the Sun, and are found in a relatively wide belt between the planets Mars and Jupiter. Occasionally they are perturbed by each other, and more regularly and dramatically by Jupiter's gravitational force. As a result, some asteroids find themselves in new orbits that cross the orbit of Mars and sometimes even that of Earth, and if we are at the wrong place at the right time then we can

[1] Douglas Adams, *So Long and Thanks for all the Fish*, Six Stories, p. 493.

get hit. (Or maybe it is at the right place at the wrong time?) Over the eons, every moon and planet has often found itself in the above situation, and has suffered the insult and injury of major impacts. You need only to look at the Moon to realize that we live in a dangerous neighborhood, and it is natural to wonder if we were struck in the past. More important is the answer to the question, "Might we be hit in the future?"

Thousands of asteroids have been discovered and there are hundreds of thousands more that are too small to be easily seen from the Earth. Our census of the largest ones is now fairly complete: there are 26 known asteroids larger than 200 km (120 miles) in diameter, the largest being Ceres, an object about one-fourth the size of the Moon, discovered in 1801 by Giuseppe Piazzi (1746–1826), a Sicilian monk.

You need only to look at the Moon to realize that we live in a dangerous neighborhood. Apollo 13 was launched on April 11, 1970. Fifty-six hours into the journey and 200 000 miles from Earth, the message "Houston, we've had a problem here" signaled the start of a frightening four-day odyssey caused by the explosion of an oxygen tank affecting critical power and life support systems. The mission had to be reprogrammed to go around the Moon and try to get back to Earth safely. This photo of the Moon taken by the Apollo 13 astronauts as they rounded it hoping to get back home shows part of the heavily cratered far-side, which we never see from Earth. The Moon always shows the same face to us because it rotates precisely once every orbit. This is not a coincidence but a consequence of the gravitational force between the Earth and the Moon. (NASA)

The gravitational force on the surface of a small asteroid, say of a diameter smaller than 100 km (60 miles) and therefore of small mass, is weak, a few hundred times less than the gravitational force on the surface of the Earth, which is your weight. If you were to visit a small asteroid and jump too hard you could easily exceed the escape velocity . . . and escape. In fact if it was very small you might have to cover your mouth when you sneeze, good manners in any case. The few small asteroids that have recently been imaged by spacecraft are not spherical, but come in a variety of strange shapes, and weak gravity is the reason why. We probably know about 98 percent of the asteroids larger than 100 km in diameter, and of those in the 10 to 100 km (6 to 60 miles) range we believe that we have cataloged about half. However, we know very few of the smaller ones, and perhaps as many as a million, potentially dangerous, 1-mile-sized asteroids may exist. Several thousand of these have orbits which bring them close to Earth, although by now you can imagine that for astronomers "close" still means many hundred thousand miles away, which for a 1-mile-size object is like saying that you are near another car when it is 1000 miles away. However, some might indeed come too close for comfort.

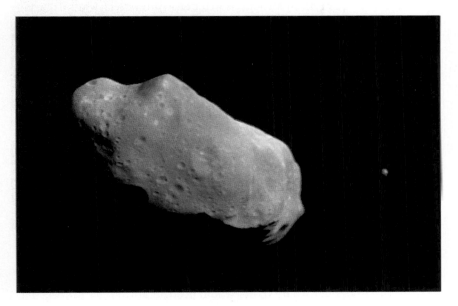

On August 28, 1993, during its six-year-long journey to Jupiter, the Galileo spacecraft took this picture of asteroid Ida, a potato-shaped object that measures about 36 miles long (60 km) and 14 miles wide (23 km). To everyone's surprise Ida has a little moon, which in this picture appears as a small dot on the right side. The tiny moon named Dactyl, the first moon of an asteroid ever discovered, is about 1 mile across. Galileo was about 6500 miles (10 500 km) from Ida when it took this picture. (NASA)

The Centaurs are a group of objects with eccentric orbits between Jupiter and Uranus. These may be the most dangerous objects in the solar system because close approaches to Jupiter, Saturn, and Uranus will inevitably perturb their paths. Chiron, discovered in 1977, has an elliptical orbit with a 50-year period that takes it from inside the orbit of Saturn to the orbit of Uranus. This 200-km diameter object, possibly a comet, will ultimately be ejected from the solar system by the gravitational influence of Jupiter and Saturn, if it is not first deflected into the inner solar system where it could then collide with a planet. That planet could be ours, a hair-raising thought. The chance of this happening in the lifetime of the solar system, however, is extremely small, but were it to happen, then the effects would be devastating.

The farthest planet from the Sun, Pluto, is at a distance of about 40 astronomical units (AU), that is, 40 times more distant from the Sun than the Earth although, because of its eccentric orbit, it goes from inside the orbit of Neptune at 30 AU all the way to 49.5 AU. However, as we have seen, our solar system does not end with Pluto. The solar system is surrounded by a vast cloud of small objects, frozen remnants from the time, 4.5 billion years ago, when our solar system formed from a dense interstellar cloud. Beyond Pluto we enter the realm of the Kuiper belt, a doughnut-shaped cloud of enormous proportions which continues to about 500 AU. It is populated by millions of small planetesimals, some several hundred kilometers in size, which is "small" only in comparison with the planets. As we saw in Chapter 3, large telescopes have recently managed to catch these very faint, elusive objects as they move slowly against the background stars. This is quite a feat, equivalent to detecting a lighted cigar in the hands of a person walking at night in New York's 42nd Street, from somewhere in Boston. Much farther out again we find the Oort cloud, a vast cloud containing billions of objects, surrounding the entire solar system and reaching out to several thousand AU.

Comets

The Kuiper belt is so far from Earth that the astronauts who traveled to the Moon in three days in the late sixties would need about 100 years to reach it. If they were to look out of the window of their spacecraft "Heart of Gold" within the Kuiper belt, they would see a black starry sky with all the familiar constellations, and an extremely bright, yellow star, maybe then seen against Orion. This star is our Sun, now very far away, but by far the brightest star in the sky. On Earth you can feel the life-giving heat from the Sun although it is far away. However, at the great distance of the Kuiper belt, very little energy arrives from the Sun and any place is a deeply frozen one, colder than anywhere we know on Earth, so cold that everything is frozen solid. If it were not for its on-board nuclear generators, obtaining their energy from radioactive plutonium, anything liquid within the spacecraft would also freeze, solar panels being of no use here. As the spacecraft entered the realm of the Kuiper belt, the astronauts would be able to observe its inhabitants: objects the size of large mountains quietly moving about the Sun. Here the gravitational pull of the Sun is over 10 000 times weaker than that which keeps the Earth in orbit about it. Consequently, these objects move very slowly, completing one orbit about the Sun in roughly several hundred years. It is not difficult for one of these objects to be deflected by another passing nearby, or by the gravitational pull of one of the giant planets. At the much more distant Oort cloud, gravitation is so much weaker again that the slightest nudge suffices to deflect an object significantly. This nudge could be caused by the passage of another star by the solar system, as we move through interstellar space in our 250-million-year-long journey about the center of the Milky Way. If pushed in the right direction, one of these frozen objects could enter the inner regions of our solar system and after some years become a memorable astronomical sight or, on very very rare occasions, cause the loss of all memories.

As it falls toward the Sun, such an object will heat up, and frozen gases and trapped dust will, slowly at first and later more rapidly, be thrown off into space leaving a beautiful tail millions of miles in length. A comet is born! The tail always points away from the Sun, pushed that way by the solar wind and radiation, and not, as many think, trailing the motion of the comet. Interplanetary space is mostly empty, so there is nothing to push the tail behind the motion, in the way that air blows the smoke of a moving diesel truck. Since dust behaves differently from gas, two distinct tails are formed. One tail shines in a shade of red by sunlight reflected by the dust, while the other is fainter, produced by the evaporating gas that fluoresces to form a bluish tail.

Comets have an irregular-shaped nucleus, a few miles in diameter, with a composition that is similar to the average composition of the solar system as a whole. This reflects their common formation out of the material of the solar nebula. Comets decay rapidly in cosmic terms, and after a few thousand years they run out of the volatile compounds which generate their characteristic tails. This transforms them into objects which look like asteroids. The Oort cloud and the Kuiper belt are the sources of new comets, about 1000 having been recorded throughout history.

Comet Halley as it appeared on March 14, 1986, in the sky over Sutherland in South Africa. Halley goes around the Sun with a period of just over 76 years and will return in 2062. (Brian Carter and Case Rijsdijk, SAAO)

Because many comets are unique, spectacular events, which appear unexpectedly, as is also the case for accidents, throughout human history they have been associated with portentous events. In most mythologies they are considered to be the harbingers of ominous events, and if you could ask a dinosaur, this view would be justified, as we shall see. Our interest in comets comes from the realization that they are made of pristine material left over from the time of formation of the solar system. They *are* the planetesimals. They contain a significant amount of cosmic dust and frozen water, carbon monoxide and dioxide, methane, ammonia and several other more complex carbon-based compounds. Ages ago they delivered these elements to the sterile Earth soon after its formation, to set in motion the complex chain of events that is the topic of this book.

In November of 1577, a bright comet appeared in the sky of Tycho Brahe, Galileo Galilei, Johannes Kepler, Giordano Bruno and Sir Francis Drake, to name just a few among 500 million, the human population of Earth in those days, less than one-tenth of today's population. Since the time of Aristotle, comets had been thought to be atmospheric phenomena, but careful observations by Brahe, and other astronomers of the epoch, led him to the conclusion that comets lay beyond the distance to the Moon. He concluded this

Launched in July of 1985, the Giotto mission was designed to study Comet Halley. The spacecraft encountered Halley on March 13, 1986, at a distance of 0.89 AU from the Sun and 0.98 AU from the Earth. This image of the dark nucleus of Halley's comet, shows spots on the surface where gas rushes away. It was taken from a distance of 12 530 miles (20 160 km). The nucleus is about $10 \times 5 \times 5$ miles ($16 \times 8 \times 8$ km) in size. Several tons of gas and dust are lost every second and in about 10 000 years the comet will have lost most of its volatile compounds. (ESA and NASA)

from the fact that the position of the comet in relation to the background stars was the same when observed from different points on Earth, something which would not be true for an object in our atmosphere since then you would be looking in different directions from different positions on Earth, and so the stars in the background of the comet would also change. These observations helped undermine the orthodox view of cosmology by then held for centuries, and they paved the way for the Copernican revolution.

Another bright comet appeared in 1682 and Edmond Halley, the English astronomer and friend of Newton, concluded that it was the same object as that observed previously in 1606. It was Halley who had convinced Newton to publish the results of his work and, based on this work, he predicted that this comet would return in 1758. Before this, it was not understood that comets could be recurring visitors, members of the solar system which moved, as Halley wrote, "in an Elliptick Orb about the Sun." The power of Newtonian physics was confirmed when the comet was indeed observed in December of 1758 by a German amateur astronomer. Unfortunately, Newton and Halley were no longer alive to enjoy the view; Newton died in 1727 and Halley in 1742. The comet was named "Halley's Comet" by the French astronomer Nicolas Luis de Lacaille (1713–1762). Comet Halley orbits the Sun in an elliptical orbit with a period of 76 years, going from 0.6 AU when it is nearest to the Sun (this closest distance to the Sun is called perihelion) to

35 AU at its farthest distance (called aphelion), beyond the orbit of Neptune. Because it is a relatively large comet, it develops a brilliant coma – a cloud of gas and dust that forms around the nucleus as it is heated – and a large, bright tail. It visited us in 1986 and it will return in 2062, to be seen by more than 9000 million people. A bad omen?

The motion of all comets and asteroids is determined by the overwhelming gravitational force of the Sun, or in the case of planetary moons by the gravitational force of their respective planets as determined by Newton's laws. However, the effect of all other objects cannot be neglected if we want to know an orbit precisely. Remember that, before Kepler and Newton, it was not really understood how planets moved. Today, we can compute the motion of any object we know in the solar system, taking into account the gravitational forces acting on it due to every other object, dominated of course by the Sun and the largest planets. This is a very complex calculation which requires the best computers available. It is also a delicate calculation, since any initial small imprecision will in time result in a very different development. Hence, the farther into the future we wish to calculate an orbit, the larger are the uncertainties. So it is that we can compute that 4179 Toutatis, an asteroid about 1 mile in size, will pass at a distance of 1 million miles from Earth on September 29, 2004, and that asteroid 1999MN will pass at a distance of "just" 500 000 miles on June 3, 2010 – not much to worry about. Asteroid 1989FC, with a diameter estimated to be 1000 feet (300 m), missed the Earth by only 450 000 miles (700 000 km) in March of 1989, less than twice the Moon's distance. It is clearly important to keep track of all these objects and to find any new ones which might pose a potential threat.

Impacts

This is not just idle talk. We recently witnessed a spectacular collision, the first large collision between two solar system bodies ever to be recorded by humans. From July 16 through July 22, 1994, pieces of Comet Shoemaker-Levy 9 collided with Jupiter. The comet consisted of at least 21 distinct fragments, with diameters estimated to be up to 2 kilometers, which plunged into Jupiter's atmosphere in a spectacular cosmic show of force. The comet had been discovered in March of 1993, and a study of its orbit had shown that it had earlier passed very close to Jupiter in July of 1992, when the strong tidal forces exerted by massive Jupiter tore it into pieces. In November 1993 there was no longer any doubt that it was on a collision course with Jupiter.

On most planets and moons with solid surfaces, and on asteroids, we see impact craters which attest to the violent history of the solar system. Objects

(a) This is a composite photo, assembled from separate images of Jupiter and comet Shoemaker-Levy 9, obtained by NASA's Hubble Space Telescope. Jupiter was at a distance of 420 million miles (670 million km) from Earth. The dark spot on the disk of Jupiter is the shadow of the inner moon Io which appears as an orange and yellow disk just to the upper right of the shadow. When the comet was observed on May 17, 1994, its train of 21 icy fragments stretched across 700 000 miles (1 million km) of space, or three times the distance between the Earth and the Moon. The apparent angular size of Jupiter relative to the comet, and its angular separation from the comet when the images were taken, have been modified for illustration purposes. (H.A. Weaver, T.E. Smith (STScI) and J.T. Trauger, R.W. Evans (JPL) and NASA STScI)

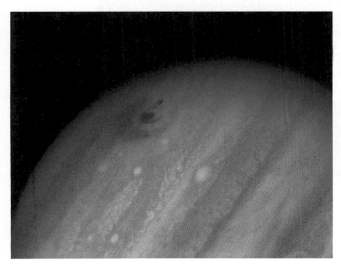

(b) This image of Jupiter, also obtained by the Hubble Space Telescope, reveals the impact site of fragment "G"of Shoemaker-Levy 9. It entered Jupiter's atmosphere from the south at a 45° angle, and the resulting ejecta appears to have been thrown back along that direction. This image was taken 1 hour and 45 minutes after fragment "G" impacted the planet. The impact has concentric rings around it, with a central dark spot 1550 miles (2500 km) in diameter and an overall size of 7460 miles (12 000 km) – about the size of the Earth. (H. Hammel, MIT and NASA STScI)

Mercury is the innermost planet and, after Pluto, the smallest planet of the solar system. Although small – of almost one-third the size and one-twentieth the mass of the Earth – it is of a density similar to that of the Earth. This image obtained by the Mariner 10 spacecraft in 1974, when it was about 3 340 000 miles (5 380 000 km) from the surface, shows the barren and heavily cratered surface of Mercury. (NASA/JPL/Caltech)

on which few craters are seen are either gaseous such as Jupiter and Saturn or, as in the case of Earth and Io, geologically active so that in time craters are erased from their surface. On Earth, erosion by wind and water can take away a few feet of a mountain every thousand years. This is imperceptible over our lifetimes but over just a few million years, really a geologic instant, a majestic 1-mile-high mountain can become just a pitiful small hill, and in much less time craters will disappear from the surface. Plate tectonics will recycle a significant part of the surface over several million years, finishing the job started by erosion. None of this happens on the Moon, or on Mercury, as both are geologically dead and without wind or water, so that on their surfaces thousands of craters have been preserved.

The fact that the craters on the Moon were the consequence of impacts, and not of volcanic origin, was discovered by the Apollo expeditions, as we saw in Chapter 3. We understand today that impacts have been an important geological and biological force, a catastrophic force originated by the tremendous energy that even a moderately sized object will deliver because of its enormous speed. This happens against a background of uniform and gradual change, so last century's argument between catastrophists and gradualists has been settled with a compromise dictated by nature. We are learning how these impacts affected Earth's history, in particular the biosphere, and wonder about how they might affect it in the future. As our Sun travels around the Milky Way it will sometimes pass close to another star, near enough to affect the planetesimals of the Oort cloud. If this happens, it could set off comet showers which will bombard our planet with horrifying consequences.

Impacts are not the only cosmic threats to life on Earth. As our Sun crosses a spiral arm of the Milky Way it might by chance find itself near a supernova explosion, or encounter a dense molecular cloud, directly affecting the environment of the solar system. These events will occur on time scales of many tens of millions of years, much less frequent than the impacts we are worrying about, and so are of interest mainly as a possible explanation for certain historical holocausts. They might have affected our planet's atmosphere and climate and brought about some of the past mass extinctions for which we have yet to find a cause. So, back to impacts.

The Earth is continuously bombarded by meteoritic dust. If you go out one night as I suggested in the preface (have you still not done so?) and wait for a while, you will see a few meteors as they burn up in the atmosphere. Every day roughly 100 tons of this material enters our atmosphere at high speed and continuously and imperceptibly settles to the ground. Some dust particles in your living-room came from space. (Think about this the next time you use the vacuum cleaner.) Some larger objects, tens of thousands of them, survive the fiery trip, occasionally being found as meteorites – most likely pieces of an asteroid. Next time you visit a science museum, go and look at these rocks from space. They might not look like much, but if you know their history they take on a special significance.

Meteorites come in a variety of forms, distinguished by their composition, and mainly related to how much iron they contain. This difference is related to their origin. Most of them are fragments of asteroids ejected after a collision, and so their composition will depend on that of the asteroid that produced them, and on its state of differentiation, that is, whether it was homogeneous or had differentiated into an iron core surrounded by lighter stony material. Some, the "irons," are just that, a heavy piece of pure iron, while the majority are "stones," because they look like stones. These are difficult to distinguish from ordinary stones unless you see them fall, or find them on the ice-covered plains of Earth's polar regions, where any stone on the surface is potentially something that fell from the sky. The concentration of the rare metal iridium in meteorites is 10 000 times higher than that of the Earth's crust, which has an almost unmeasurable concentration of this element. This is because iridium readily alloys with iron, and so, as the Earth differentiated, this rare element collected with iron as the core formed.

The remains of approximately 150 impact craters have been identified on our planet. More craters have been identified in Australia, North America, and eastern Europe because these areas have been relatively stable for very long geological periods, thus preserving the geological record. The first impact crater to be recognized as such is the well-known Barringer Meteor Crater located near Winslow, Arizona. It is not a very large one as craters go, with a diameter of about 1 mile (1.6 km) and a depth of about 600 feet (200 meters). For a long time, the origin of this crater was the subject of controversy, some thinking that it was volcanic. The discovery in the 1920s of fragments of the Canyon Diablo iron meteorite, which had been found nearby, within the deposits that partially fill the structure, proved its impact origin. A range of minerals which are only formed by the extremely high pressures and temperatures generated during an impact, such as shocked quartz and tektites, further confirmed its origin. Tektites are glass beads formed as quartz grains are instantly melted by the extreme heat produced on impact and then cool to form glass as they fall back to Earth. The crater's age has been estimated at 50 000 years, making it the most recent crater of its size on Earth. One of the largest known craters found on the surface of the Earth is the one at Manicouagan in Quebec, Canada. This crater, 40 miles (70 km)

A silent witness to cosmic violence on Earth, this crater in Arizona is the result of the impact by a relatively small projectile that struck the Earth about 50 000 years ago. It has a diameter of about 1 mile and a depth of 600 feet. (NASA Lunar and Planetary Institute)

in diameter, was produced about 215 million years ago in the late Triassic period.

The largest known crater, the Chicxulub crater in the northern part of the Yucatan peninsula of Mexico, is about 125 miles (200 km) in diameter with a depth estimated to be 8 miles (13 km). Its center is near the town of Puerto Chicxulub on the coast of the Gulf of Mexico. It is, however, buried deep under several hundred meters of sediment and therefore hidden from view. It was discovered in 1950 after geologists working for Pemex (Petróleos Mexicanos), searching for oil deposits, found a large underground circular feature centered at Chicxulub. At that time, they did not identify it as an impact crater. As we shall see, it is a unique find, because it tells us an almost incredible story about life on Earth.

Megatons

To produce a large crater, a colossal quantity of energy is needed. We can measure this energy in a unit called a megaton (1 Mt), which is equivalent to the explosive energy of 1 million tons of the chemical trinitrotoluene, commonly known as TNT. You would need a 400-mile-long freight train to transport this much TNT. The energy (E) of a projectile can easily be obtained

The Manicouagan reservoir, located in a rugged, heavily timbered area of the Canadian Shield in Quebec Province, is a large annular lake 60 miles (100 km) wide, seen on this winter photo taken from the Space Shuttle. It marks the site of an impact by a large meteorite which happened about 215 million years ago. The crater has been worn down by many advances and retreats of glaciers and other processes of erosion. (NASA)

if we know its mass (M) and its velocity (v) by multiplying the mass by the square of the velocity and then dividing by two: $E = \frac{1}{2} M \times v^2$. The velocity is the most important factor since the energy depends on its square. Thus, a car collision at 60 miles per hour will be four times as damaging as one at 30 miles per hour, because if you double the velocity, the energy increases by a factor of 4, the square of 2.

The escape velocity, mentioned previously in Chapter 3 in relation to the escape of atmospheric gases, is also the velocity reached by any object which starts from rest very far away and falls onto the surface of Earth. This is the velocity before the object enters our atmosphere since, if it is small, it will

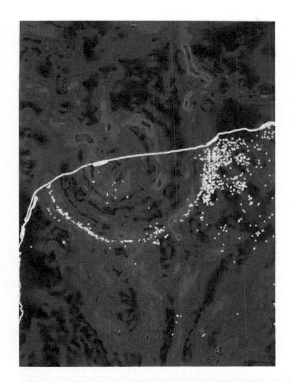

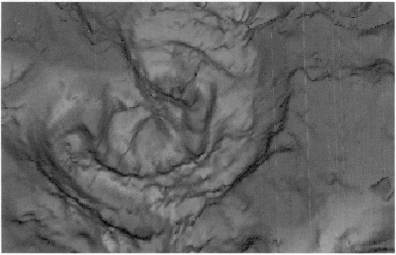

These images show maps of the gravity anomaly – local changes in gravitation due to variations in the underground density – on the northwest coast (marked with a white line) of the Yucatan peninsula of Mexico. The circular feature is the Chicxulub crater, which is buried under several hundred meters of limestone, near the town of Puerto Chicxulub, not far from Mérida. North is to the top. (The Geological Survey of Canada, Natural Resources Canada)

then be slowed, by atmospheric friction, which will also heat it up. The outcome depends on the size and constitution of the incoming object. It is also a *minimum* velocity because most objects, such as comets or asteroids, do not start from rest but have some initial velocity, and will hit Earth with more than the escape velocity of 11 km/s which, as we have seen, is enormous: about 25 000 miles per hour. Let us consider a small, 50-meter (160-feet) diameter projectile. It will have a mass of about 150 000 tons if made out of stone, and about three times this if composed of iron. Our projectile, arriving with at least the escape velocity, will be carrying an energy of at least 1 Mt, and this could easily be 15 Mt if the velocity were higher and it were heavier. The point is that even a smallish object will carry a vast quantity of energy, and this can wreak havoc upon impact.

The word "megaton" brings to mind a tragic episode in the history of *Homo sapiens*. Over half a century ago, on August 6, 1945, the sky over Hiroshima, a city in the southwest of Japan, was partly cloudy and the weather was good. People were starting the day as they usually did in those difficult times of war. This was the Second World War, a sad chapter in our recent history, which to me almost proves that we are not very intelligent, or that if this was the product of human intelligence, then we surely need something different.

Early that day, a flash of bright light accompanied by scorching heat, followed almost immediately by the sound and force of a powerful blast wave, changed everything. Over a distance of a few miles, everyone was killed almost instantly, and everything destroyed. For 100 000 people, time ended at 8:16 that morning. Countless others were horribly injured, and the city was erased from the landscape. As the dust settled, all that was left were the charred remains of buildings and bodies, and the harrowing sounds of those still in agony. All this ghastly horror was caused by a rudimentary atomic bomb, innocently named "Little Boy," with the explosive power equivalent to that of just 15 000 tons (15 kilotons) of TNT. It would take 65 of these bombs to make a megaton.

Today, many nations have nuclear weapons, some more than 1000 times as powerful as the one dropped on Hiroshima, or the one detonated a few days later, on August 9, 1945, over the city of Nagasaki, with the same tragic result. Many thousands of megatons are stockpiled in several countries, mostly in the US and in the former Soviet Union, we hope never to be used. Beyond the effects of a large meteoritic impact, or a conventional chemical bomb for that matter, a nuclear explosion also generates a large quantity of lethal nuclear radiation, and leaves behind radioactive isotopes which over a long time, depending on their half-lives, continue to do harm.

A symbol of *Homo sapiens*'s folly. For 100 000 people in Hiroshima, time ended at 8:16 on the morning of August 6, 1945, as recorded on the watch of one of the victims, Kengo Futagawa, who was crossing a bridge on his way to do fire prevention work 1600 yards from the hypocenter. He jumped into the river, terribly burned, and died on August 22. (Main image US Army; smaller image Hiromi Tsuchida and the Hiroshima Peace Memorial Museum)

Earth's atmosphere protects us from the multitude of small projectiles, the sizes of grains of sand or small pebbles, which pelt our planet every day. Hundreds of tons of material from space arrive on Earth in this way every day. The meteors in our night sky – we call them shooting stars – are visible signatures of bodies of this type burning up high in the atmosphere. Although small, they travel at such high speeds that a tiny grain can carry more energy than a 22-caliber bullet. Without our atmospheric shield they could be lethal if they were to hit you. Fortunately, up to a diameter of about 10 meters most stony meteoroids are destroyed in the atmosphere in a terminal explosion triggered by the heat of friction. An iron meteorite will survive its fiery descent much better than a stone.

Earth's atmosphere cannot significantly slow down objects a few tens of

Even though most meteoroids are small, ranging in size from dust grains to small pebbles, their entry into our atmosphere at enormous velocities can generate sufficient energy to ionize the meteoroid material and gases in the atmosphere producing a flash of light – a meteor or a "shooting star." This meteor was observed as part of the Leonid meteor shower, a yearly event occurring about November 17 when the Earth crosses a meteoroid stream generated by comet Tempel-Tuttle. To the left of the meteor you can see part of Orion, and at the right edge of the image you can recognize the Pleiades. Bright Aldebaran is visible above the Hyades star cluster. (Arne Danielsen)

meters in diameter or larger, and these can strike the ground or the ocean at high velocity doing considerable damage. Fortunately, these objects are much less common. Meteor crater was formed by a chunk of iron, perhaps 35 meters in diameter: the Canyon Diablo meteorite that, as we have seen, fell about 50 000 years ago. Anything living nearby would have been instantly killed by the explosion that produced this crater, estimated to have been equivalent to the explosion of a several megaton atomic bomb.

Early in the morning of June 30, 1908, a cosmic missile thought to be about 160 feet (50 m) in size, comparable to a building, entered the atmosphere at a high velocity, heating up as it fell and creating a streaking fireball as bright as the Sun. Upon hitting the high-density region of our atmosphere at a height of about 3 miles (5 km), it exploded above the desolate Tungus river region of western Siberia, generating a powerful blast estimated to have been equivalent to a 10-megaton nuclear explosion. This was almost 700 times more energetic than the bomb that annihilated Hiroshima. It is said that the sonic boom was heard as far away as England, and that the night sky over Europe was uncommonly bright for several days afterwards, although of course nobody knew then what had happened. Subsequent expeditions into this remote region found thousands of fallen trees pointing radially outward from the center of the explosion, many broken-off as if they had been mere matchsticks. The explosion devastated the surrounding forest within about a 20 miles (30 km) radius. A few eyewitnesses, fur tradesmen and Tungus herdsmen, miles away from the explosion, later recalled being hurled through the air and falling unconscious. The ground shook, the wind roared, and animals and trees were scorched. Anyone closer to the center of the explosion would have been instantly cremated without first dying. Luckily for most people then alive (we were 1.75 billion then) this all happened in a remote and sparsely populated region, so remote that it was only 19 years later, after a grueling trip, that the first scientific expedition, led by the Russian Leonid Kulik (1883–1942), reached the site.

Leonid Kulik was born in Tartu, Estonia, and studied in St. Petersburg (later Leningrad). In 1920 he took a position at the mineralogical museum in St. Petersburg where he devoted his time to the study and the acquisition of meteorites. He found out about the Tunguska event from a reprint of an old newspaper article and thus began his quest to find the site of impact and recover meteoritic material. His first expedition in 1921 was unable to reach the remote site, but in 1927, in spite of opposition by his skeptical colleagues, he managed to obtain funds to mount a second expedition. After several months of difficult march across the swampy terrain, navigating rivers on rafts, and crossing the snow-covered landscape with the help of horse-drawn

(a) This photo taken by Leonid Kulik in 1927 shows the scene of utter devastation he found. For many miles he found fallen and burned trees all pointing away from the blast. (University of Bologna www-th.bo.infn.it/tunguska)

sleds he finally reached the site of the explosion. "The results of even a cursory examination exceeded all the tales of eyewitnesses and my wildest expectations," he wrote.[2]

On his return to Leningrad his photographs convinced his colleagues that something awesome had occurred at Tunguska. Two additional expeditions led by Kulik, one in 1929 and the other in 1938, reached and explored the site but he never found the sought-after meteorite. In 1942, Kulik, then fighting against the German invaders, was captured and imprisoned. He died of typhoid fever on April 24, 1942.

[2] Quoted in "Virtual Exploration Society" website, http://www.unmuseum.com/kulik.htm

(b) A portion of one of the photos from Kulik's aerial photographic survey (1938) of the Tunguska region showing a great number of trees all fallen in the same direction. (University of Bologna www-th.bo.infn.it/tunguska)

Had this event occurred in a populated area, the devastation would have been immense, a great natural disaster worse than the most powerful earthquake. Had asteroid 1989FC struck in 1989, it would have delivered an estimated 100 megatons, 7000 times more powerful than the Hiroshima bomb, a scary thought.

We can obtain an estimate of the occurrence of large impacts on Earth by studying the craters we see on the volcanic marias of the Moon which formed more recently than the rest of the lunar surface. As determined from the study of the Apollo samples returned to Earth, these maria formed about 3.5 billion years ago after the heavy initial bombardment of the Moon was over. The results of these studies suggest that the Earth is struck by a projectile producing a crater larger than 15 miles (25 km) about once in every 10 million years. Smaller objects are more abundant than the rare larger ones which create the most damage. It is estimated that one Tunguska type object will strike the Earth every 500 years, and objects of 6 miles (10 km) diameter, such as that which produced Chicxulub, will hit once every 100 million years.

Perhaps the greatest danger comes from objects of intermediate size, say 0.5 mile in size, large enough to cause significant damage yet small enough to be numerous. Such objects will strike the Earth every 100 000 years on average, releasing tens of thousands of megatons. However, their arrivals are

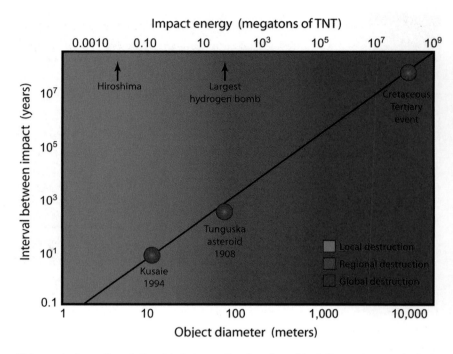

This graph shows the relationship between the size of an object, its energy upon impact, and the expected interval between impacts. Note that the scales are *logarithmic*, meaning that every tic-mark on the scales corresponds to a change by a factor of ten or 100. A 10-meter diameter object, such as the one that caused an atmospheric explosion over the South Pacific island of Kusaie in 1994, delivers about 0.1 megaton and is expected to occur every ten years on average. We do not worry much about objects of this size, although they carry more energy than the Hiroshima bomb, because they will most likely explode in the atmosphere and not reach the ground. Also plotted are the Tunguska and Chicxulub objects. Note that an object of about 1000 meters in diameter (1 kilometer, about 0.6 mile) can deliver about 100 000 megatons and will occur every 100 000 years. We need to find all of these, but also the smaller and more abundant objects that could cause great damage if they were to hit a populated area. (José F. Salgado after David Jewitt)

random, unpredictable events which follow no timetable. When we say that a random event occurs once every 500 years, what we mean is that on average this is so, but not that every 500 years it will happen. We cannot say that the next Tunguska will happen in 2408; it could be 1000 years before the next event or it could be tomorrow morning. We simply do not know with certainty. For a 0.5-mile-size object, what it means is that within any year the odds of being struck are 1 in 100 000. You probably have bought lottery tickets at some time, hoping for a win, where the odds were much worse than 1 in 100 000. Should we win this cosmic lottery, we lose!

There is discussion about the details of all this, whether our estimates are too high, or maybe too low, and we shall learn a lot from additional exploration. We might also discuss what to do about a collision but there is no discussion about the fact that some day we will be hit by a large projectile. There is no need to lose any sleep about this; right now no asteroid or comet is known to be on a collision course with the Earth. Nevertheless, we have no way of knowing if a mountain-sized planetesimal is right now being deflected into a trajectory which will, in a few hundred years time, bring it to an encounter with Earth.

If we were to have sufficient warning, the incoming object could be deflected or destroyed by a nuclear device of several megatons, as some have advocated. It might be the only redeeming value of our nuclear weaponry. However, this is quite difficult in practice. Although it would be a powerful blast, it is small compared with the energy carried by the asteroid or comet, and so would cause only a tiny effect. It would alter the velocity of the object by a small amount, and this could, over a long time, perhaps several years, change its orbit sufficiently for it to miss Earth. If it is a well-known asteroid with a precisely determined orbit, then this might be feasible. However, if it is a new comet or a smallish, previously unknown asteroid there might be insufficient time for this. Depending on its nature, it might be beyond the orbit of Jupiter and very faint. It would be difficult at such great distance to discover it, and to measure an orbit with the necessary precision to be able to tell if it would, beyond any doubt, hit us. If we did not know this precisely, we might make things worse by trying to deflect it. "Uhmm . . . We have a problem, guys, we seem to have pushed it in the wrong direction . . ." On the other hand, if we discovered the object when it was already near, and therefore bright, say at a distance of about 100 times the distance to the Moon, it would then hit us in about two weeks. It would, by this time, be relatively bright, but it would also be coming straight at us. That is, it would show very little motion across the sky to be recognized as a moving object against the background stars, which is how astronomers usually recognize these things in the first place. So we might not detect it until it was too late anyway.

Astronomers are conducting increasingly sensitive searches for all of the larger asteroids that could become dangerous. The powerful Arecibo radar could be used to measure accurately the trajectory of such an object, to find out whether it was on a collision course with the Earth, although it cannot see the entire sky. However, as we have seen, there is no guarantee that we will find all such threats in sufficient time. Furthermore, one could question whether having a set of missiles armed with nuclear warheads, always ready, might not pose a greater threat to life on Earth than the once-in-a-few-thousand-years "missile" against which we want to defend ourselves. Realistically, all we might be able to do is to duck, although it is likely that this would not help either. As we shall see in Chapter 8, other more urgent if less spectacular threats are waiting with much more certainty, and will in all likelihood cause our exodus before any cosmic projectile, such as the one which hit the Earth 65 million years ago, strikes.

Chicxulub

It was a day on Earth like so many others in the Cretaceous period, a day of only 22 hours in a world that was not quite as it is now, and where the continents were not quite where they are now. It was a sad forgotten day, only to be remembered 65 million years later by the descendants of those who survived the nightmarish event. On the previous night, as on several nights before that, a brilliant comet had been visible, its luminous tail spanning the entire sky. Creatures of the night, accustomed to the cycles of the Moon, were confused by this added light. The afternoon had been cloudy, with a warm humid breeze sweeping the tropical land, lightly agitating the forest of palm trees and large ferns through which giant dinosaurs were walking toward the coast. Small mammals scurried between stones avoiding being trampled by *Tyrannosaurus rex*. A huge flying Pterosaur, a *Quetzalcoatlus*, with a 10-meter (33-feet) wingspan, larger than that of a small plane, swooped from a tall ledge and aimed its long beak at a small school of fish swimming in the shallow coastal waters.

An afternoon thunderstorm had just passed, and a beautiful rainbow was painted on the sky. This multicolored arch is the result of raindrops acting like millions of tiny prisms, breaking up the light from the Sun into its component colors, and then projecting this onto the opposite part of the sky. There is supposed to be a pot of gold at the end of the rainbow, a frustrating symbol of a better but unreachable future, since you can never get there no matter how long you travel. At the end of *this* rainbow, however, there was the blackness of death. The brilliant comet from the previous night was now so bright that even its tail was visible in daylight, and its brightness increased very rapidly. The incoming object, the size of a mountain, took only 3 hours to cross the vast space between the Moon and the Earth. It took less than 1 second to traverse the Earth's atmosphere and slam into the ground, a period so short that nobody would have had time to realize what had happened, even had they possessed the brains to think about this. The loudest of sonic booms, the chorus of a million thunders, the last thing to be heard by many, signaled the event and as it subsided it left behind a brutal wave of terror and death which encircled the globe.

The 6-mile (10 km)-sized object slammed into the Earth with the energy of 100 million megatons, piercing the planet's crust and producing a crater 8 miles (13 km) deep. The comet or asteroid, we really do not know which it was, vaporized completely and kicked up billions of tons, more than 200 000 cubic miles of vaporized material, from the Earth's crust high into the atmosphere. *It did so at Chicxulub.* A giant mile-high wave, the mother of all tsunamis, traversed the Gulf of Mexico destroying all it encountered as it hit the coast of North America and the Caribbean islands, reaching as far as today's Hispaniola and Texas. The dust, soot, and smoke in the atmosphere blocked out all sunlight for several months, or even years, and caused global temperatures to plunge to near freezing. Later, temperatures rose to high values due to the increase of CO_2 injected into the atmosphere. This double blow led to a collapse of the ecosystem with devastating consequences for all life on Earth. Photosynthesis stopped, plants died, and those animals that fed on plants, or on other animals, starved. At Chicxulub, the rocks contained a high proportion of sulfur which was injected into the atmosphere by the impact and later formed sulfuric acid which rained onto the surface, adding insult to injury. About three-quarters of the species on Earth, including the dinosaurs, suddenly disappeared in what is called the "Cretaceous–Tertiary extinction." When the dust settled, it did so in a thin layer all around the globe. Since it was composed of vaporized material from the projectile, it contained a much higher fraction of the rare element iridium than is common in the Earth's crust.

Geologist Walter Alvarez and his father, Nobel Prize winning physicist Luis Alvarez, were studying the thin layer of clay at the Cretaceous–Tertiary (K–T) boundary in the mountains near the medieval Italian town of Gubbio in the 1970s. (By the way, the "K" for Cretaceous comes from the German name for chalk, "Kreide," and is used so as not to confuse Cretaceous with Cambrian.) The Alvarez team wanted to learn how much time this layer represented, since this would be important toward understanding this mysterious mass extinction, so clearly visible in the fossil record, as one goes from Cretaceous sediment through the inch-thick clay boundary into Tertiary sediment. The question they sought to answer was: did this mass extinction happen suddenly, and therefore catastrophically, or was it a gradual process?

Imagine that you drop red ink at a constant rate, say one drop per minute, into a large river and that you sample the water once a day at a station a few miles downstream. If the flow of water is high, the ink will be

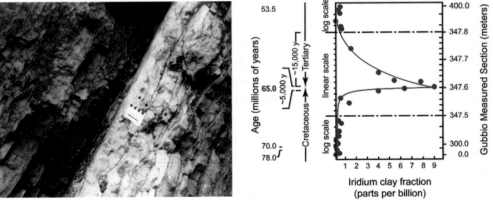

A thin exposed bed of clay dated at 65 million years separates Cretaceous (white) from Tertiary (brown) sediments. In this photo of the boundary at Gubbio, Italy, the older sediments are toward the bottom right. The iridium layer is the thin dark layer (running diagonally) over the Cretaceous rock. To determine if the mass extinction at the K–T boundary had been a sudden catastrophe or a slow process the iridium concentration was measured. This is a difficult and delicate measurement using the technique of neutron activation analysis, which can determine concentrations of parts per billion. The result is shown in the graph adapted from the *Science* paper by Walter Alvarez and colleagues. The points mark the iridium concentration measured at various positions across the boundary (you can see the boreholes on the enlargement). The value found at the boundary was so high that it led to the impact idea, where a large quantity of iridium is deposited in a short time. Confirming this idea was the finding that in many areas of the world the layer also contains soot from fires and small grains of shocked quartz, evidence of the high temperatures and pressures generated by the impact. (Photos courtesy of Kosei E. Yamaguchi, graph by José Salgado after Walter Alvarez)

diluted to a greater extent than if, as during a dry season, the flow is low. Over the years you would be able to reconstruct the history of water flow by measuring the concentration of red ink. You might say that a drop is such a minute fraction of all the water flowing, maybe just one part in 1000 million (called one part per billion, ppb, which is represented by 1 second in 32 years), that measuring it is impossible, and this might be true for the case of ink in water. However, using advanced techniques it is possible to measure the concentration of certain substances at the level of ppb, and this is what the Alvarez team set out to do with iridium in the clay deposits of the K–T boundary.

They reasoned that if meteoritic dust settled at approximately a constant rate, as with the ink of our example, and if they could detect the concentration of iridium in the sediments, the water of our example, this would provide them with a natural clock. A high concentration of iridium would mean slow deposition of sediment allowing for a larger quantity of iridium; a low concentration would mean a fast deposition. They were startled by the outcome. Iridium in the layer, although less than 10 ppb, was 100 times more abundant than in the surrounding deposits. This did not fit at all with the idea of a constant rate of deposition from meteoritic dust. There was simply too much of it. The measurements of this "iridium anomaly" at the K–T boundary at many other locations throughout the world showed that this was not merely a local effect, and led to the conclusion that it was the result of the impact of an extraterrestrial projectile with a size of about 10 km.

A paper published in the prestigious journal *Science* in 1980, entitled "Extraterrestrial cause for the Cretaceous–Tertiary extinction" and written by Luis and Walter Alvarez, Frank Asaro and Helen Michel, presented the results and set the stage for a great scientific debate which still goes on in some quarters. However, many years of careful research have corroborated this result and provided further evidence that a great impact definitely did occur 65 million years ago. The discovery ten years later of the impact nature of the Chicxulub crater, and most significantly the determination of its age as 65 million years, obtained from radioactive dating of excavated minerals, left little doubt about the relation between this colossal cosmic event in the history of Earth and the extinction of almost all life.

Thus, Lady Luck has played a much larger role in the history of life than we could have imagined only a few years ago. There is nothing more unlucky, should you be a dinosaur, than being wiped out by the freak impact by a piece of cosmic leftover. We, the mammals, might see this as a stroke of good fortune, an accident that ushered in a new age – *our* age.

Chapter 7

Other worlds

Ursa Minor Beta is, some say, one of the most appalling places in
the known Universe[1]

Are we the result of a most improbable cosmic accident, the lonely inhabitants of this immense Universe? Or is life a common occurrence, a cosmic
imperative, arising anywhere that conditions are favorable? We would
dearly like to know.

Giordano

Four hundred years ago, at a place in Rome called Campo dei Fiori, on
February 17 of the year 1600 to be precise – almost 1600 years after a 33-year-
old man had been nailed to a cross and left to die because the Roman authorities disliked his views – a 52-year-old man was stripped, bound to a stake,
and burned alive by order of the Pope, Clement VIII, for being an "impeni-
tent, stubborn and obstinate" heretic. This man had written: "Innumerable
Suns exist, innumerable Earths revolve about these Suns in a manner
similar to the way the seven planets revolve around the Sun. Living beings
inhabit these worlds." Clearly a great dose of heresies. In 1593, Giordano
Bruno, the author of these lines, a contemporary of Tycho and Galileo, was
imprisoned in Rome by the Inquisition for holding these heretical views.
Bruno was born in Nola, near Naples, Italy, in 1548 as Filippo Bruno. He
became Giordano upon entering a Neapolitan cloister of the Dominican
Order in 1563, at age 15. He was ordained as a priest in 1572. In 1576 he fled
to Rome after getting into trouble with his superiors on theological matters,
and a process was started against him. He left dangerous Italy for a tumul-
tuous life, spending time in France, England and Germany, always on the
move, lecturing and writing on philosophy, magic, astronomy, and the art
of memory. This is how he described himself in a letter asking for permission to teach at Oxford University in 1583:

[1] Douglas Adams, *The Restaurant at the End of the Universe*, Six Stories, p. 169.

To the most excellent the Vice-Chancellor of the University of Oxford, its most famous Doctors and celebrated Masters – Salutation from Philotheus Jordanus Brunus of Nola, Doctor of a more scientific theology, professor of a more purer and less harmful learning, known in the chief universities of Europe, a philosopher approved and honourably received, a stranger with none but the uncivilized and ignoble, a wakener of sleeping minds, tamer of presumptuous and obstinate ignorance, who in all respects professes a general love of man, and cares not for the Italian more than for the Briton, male more than female, the mitre more than the crown, the toga more than the coat of mail, the cowled more than the uncowled; but loves him who in intercourse is the more peaceable, polite, friendly and useful – Brunus whom only propagators of folly and hypocrites detest, whom the honourable and studious love, whom noble minds applaud.[2]

In 1591, at the invitation of the Venetian patrician Giovanni Moncenigo, who wished to learn his art of memory, Bruno returned to Italy, which he had left fourteen years earlier as a refugee, only to be betrayed by Moncenigo who, dissatisfied with his teachings, denounced him to the Holy Office. After seven years of imprisonment and nights of torture did not sway him from his convictions, and refusing to recant, Friar Giordano was handed to the secular Roman authorities with instructions that he be punished "with as great clemency as possible, and without effusion of blood" – a euphemism for burning at the stake. To his judges he said "Greater perhaps is your fear in pronouncing my sentence than mine in hearing it."

His view on the plurality of worlds was not in accord with the ideas about the Universe and our place in it, affirmed by the ecclesiastic authorities of the time, and so in the name of God they burned him at the stake. Thus ended Bruno's life, unnoticed but by a few, a martyr for freedom of thought. A restless life of struggle against dogma and prejudice, striving to understand our world through the senses and through reason, rejecting authority and convention, ended tragically. This was neither the first nor last time that this happened in the history of mankind. We have come a long way in the 400 years since then, though with sometimes disturbing lapses. We are fortunate enough to live at a time when we think that the most important scientific results are precisely those which go against dogma of any kind, and that we need not fear dire consequences should our discoveries not agree with established theory. I hope we can keep it this way, although we know of countries where people still seem to live in the 1600s. Science only

---

[2] Quoted in J. Lewis McIntyre, *Giordano Bruno – Mystic Martyr*, Kessinger Publishing Co., Montana, 1903.

requires rigorous proof and independent confirmation of any new result for it to be accepted. This process can sometimes take a long time, simply because, contrary to what you may think, the facts do not necessarily speak for themselves. Interpretation is necessary, sometimes straightforward, but at other times quite involved and dependent on aspects of theory, before the facts take on meaning. *So, was Giordano Bruno right or are we alone?*

Aliens

As mentioned earlier, a great step toward understanding the origin of life would be the discovery of life, any life, on another world. A bacterium would do, even a fossil one would be great. It would be the discovery of the century, and as such would have to be carefully scrutinized, double and triple checked, and then checked again before being accepted as fact. Most scientists, however, would not be surprised by such a discovery. It is reasonable considering what we have learned about the formation of the solar system and the evolution of life. The discovery of *intelligent* life elsewhere (we are generously assuming that it exists on Earth) is an entirely different matter, and it would possibly qualify as the greatest discovery of all time. Here we are not so confident concerning what to expect, and a good fraction of scientists are quite skeptical about this possibility. However, "the proof of the pudding is in the eating" and so searching for intelligent life, no matter how unlikely we might think this is, is nevertheless a worthwhile endeavor.

Some people insist that aliens are here already: they say that they have had close encounters with all sorts of beings from other worlds. They claim that they were abducted by, or went voluntarily with, beings commonly described to be like those in the latest Hollywood production. The aliens that they claim to have seen look very much like modified humans, with large eyes and heads, which points to them being figments of the imagination – and not a very creative one at that. Looking at the diversity of life on Earth and thinking about how it has evolved should convince anyone that any aliens will have as much resemblance to us as a doorknob. Some people even offer us the knowledge they gained in one of these encounters with aliens, who, they claim, are so much more advanced than us (the reason for the large heads) that they have discovered the secrets of immortality, beauty, and other good things. They are willing to let you in on these secrets, usually for a modest fee of course. A surprisingly large number of people are quite willing to believe all this, and will take denials by government officials and scientists (including me) as a confirmation of their suspicions, since it is understood that the government will keep such things secret. Why this would be so escapes me, but it fits with the theme of somber conspiracies, a great subject for movies (as in the film *Men in Black*). Well, enough of that, most of the time no harm is done, and it is good for a few laughs. Unfortunately, too often, in the hands of unscrupulous merchants, this entertainment becomes a tool to exploit those who have no easy way to know the facts, and are led to confuse reality with fiction. With the illusion to find an easy way out of their unhappy condition, or upon the news of impending doom, they will act foolishly, jeopardizing their properties and sustenance and sometimes even endangering their lives or those of others.

One wonders why so many persons are willing to believe that aliens are amongst us, that at the Arecibo Observatory we talk with "them" all the time, that these highly developed life forms, after completing a fantastic journey with inconceivable technology, crash miserably on some lonely desert. It cannot be explained simply as ignorance of scientific facts; there is more to it. Perhaps it is that, for many, the extraterrestrials are a modern incarnation of the ancient gods that we no longer need to push the planets in their orbits. They provide comfort and hope in an uncomfortable and hopeless world. They might even take us with them to another, better world. From a scientific point of view all this does not stand the test of rigorous proof and independent confirmation, and deserves attention only as a social phenomenon.

The first place to search for signs of life is, of course, in our solar system. The best place would be one where water was present in the past or is there now, since this is the element of life. Could there be life, past or present, on one of the other planets or on any of the larger moons? Starting in the 1960s, a remarkable series of space probes visited most of the planets in our solar system. These investigations confirmed that currently none of these nearby worlds seem suitable for complex life forms.

Martians and Venusians

Mars, long regarded as the most likely planet to harbor cosmic company, has a cold, dry, and apparently sterile surface, though it may have harbored primitive life long ago. It is a small planet with only one-tenth the mass of the Earth and half its size. It is 50 percent farther from the Sun than the Earth is. It has a very thin atmosphere, mostly composed of carbon dioxide and nitrogen, with the pressure at the surface 100 times less than it is on Earth. However, there is enough of an atmosphere for Mars to have clouds, and winds which create dust storms and seasonal changes.

Mars has been the subject of hundreds of science fiction novels and many

I found this hanging from a cliff outside the gate of the Arecibo Observatory. The Spanish text at the bottom says: "here" and "now," the prevalent view by a large fraction of the population. (Tony Acevedo)

movies initially inspired by the publication in 1877 of descriptions of Mars by the Italian astronomer, Giovanni Schiaparelli (1835–1910). He observed from Milan, Italy, and described the surface of Mars as showing channels: "canali" in Italian. This was translated as canals, therefore implying that some form of technology existed on Mars. This theme was taken up by the American Percival Lowell (1855–1916), who founded the Lowell Observatory in Flagstaff, Arizona, to study Mars. (It was here that Clyde Tombaugh (1906–1997), an observatory staff assistant, discovered Pluto on February 18, 1930 after searching for many years. He had examined over 90 million star images on photographic plates finally to find one that seemed to move against the background stars.) Over many years Lowell popularized the idea that there were Martians living on Mars. But there are not, and the canals were just artifacts of not seeing the surface clearly, combined with some measure of wishful thinking. However, the real surface of Mars is extremely interesting with Olympus Mons, the largest (extinct) volcano in the solar system, and Valles Marineris, a canyon of enormous proportions resembling a gigantic wound on the body of Mars. Most relevant to this story, Mars has what may be beds of ancient rivers, suggesting that water once flowed on its surface. High-resolution images obtained by the Mars Global Surveyor spacecraft, in orbit about Mars, show what appear to be gullies carved out by water present under the Martian surface. That water might be currently present in underground aquifers is a very exciting possibility.

Mars was much more like Earth when it formed out of the solar nebula, but its different location and much lower mass meant quite a different development from that of the Earth. Furthermore, because Mars has only two puny moons (Phobos and Deimos), its axis of rotation is not stabilized like the Earth's, and has in the past changed direction by large and unpredictable amounts in response to the gravitational influence of the other planets. This would have had a dramatic effect on the development of any life on Mars, had it originated there. However, as we have seen, life arose on Earth very early, and the question everyone would like an answer to is: did the same thing happen on the young Mars? A positive answer would tell us that life originates easily under the right conditions and is therefore quite common in the Universe. This would be an extremely important result, and is the reason why the announcement in 1996 that what appear to be traces of an ancient microorganism had been found inside a 3.6-billion-year-old Martian meteorite named ALH84001 (the first meteorite found in 1984 near the Allen Hills region), found in Antarctica, created such great interest. Later, chains of microscopic magnetic crystals of iron oxide, called magnetite, were found inside the meteorite, undistinguishable from those found inside certain "magnetotactic" bacteria which use these to orient themselves. The evidence is, however, far from conclusive, and although the headlines heralded "Life on Mars," most in the scientific community are quite skeptical for two good reasons. First, it is well known that "organic" materials, such as those found in the meteorite, can be produced by non-biological processes. Second, the putative fossils are 1 million times smaller than known bacteria, so small that it would be hard to fit into them the complex molecular machinery necessary for life. This extraordinary claim, as happens with all such claims in science, will need to be corroborated by further studies, especially on the surface of Mars. Finding the remains of

Valles Marineris is a canyon complex of enormous proportions resembling a gigantic wound on the body of Mars. It was named after the Mariner 9 spacecraft that took the first images of the canyon in 1971. The Viking Orbiter and Lander missions arrived at Mars in 1976 and took the 102 images used for this mosaic, a view similar to what you would see from a spacecraft at a distance of 1500 miles (2500 km) from the surface. Valles Marineris crosses the equatorial region of Mars for a length of roughly 2000 miles, one-fifth of Mars's circumference. The canyon is 300 miles (500 km) wide and in some places reaches a depth of 5 miles. For comparison the Earth's Grand Canyon is only 300 miles long and 1 mile deep. The three round dark spots to the left are extinct volcanoes in the Tharsis region of Mars. At the eastern (left) end of the canyon are what appear to be dried-up channels left by water erosion. (NASA)

 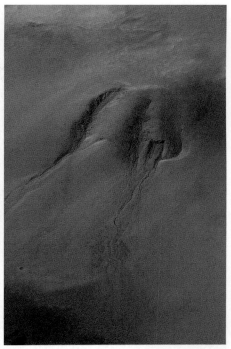

Many places on the surface of Mars show what seem to be the beds of ancient rivers, suggesting that water once flowed on its surface. The high-resolution image on the right, obtained by NASA's Mars Global Surveyor in 2000, covers an area approximately 1.9 miles (3 km) wide by 4.1 miles (6.7 km) high and is part of the south-facing wall of a crater in the Noachis Terra region of Mars. The image on the left shows this crater which is approximately 12 miles (20 km) in diameter. The image was obtained by the Viking 1 orbiter in 1980. The large mound on the floor of the crater is a sand dune field. In the high-resolution view you can see channels and associated aprons of debris that are interpreted to have been formed by groundwater seepage and surface runoff. The lack of small craters superimposed on the channels and apron deposits indicates that these features are geologically young. It is possible that these gullies indicate that liquid water is present within the Martian subsurface today. (NASA/JPL/Malin Space Science Systems)

past life on its surface, say something like a fossil stromatolyte, would bring us much closer to understanding the origin of life. And if life did not develop on early Mars we would very much like to know why.

Finding life anywhere in the solar system would provide answers to such important questions as to whether the genetic code is really universal, following the rules of yet-to-be-discovered natural laws, or if it is simply an accident. This would tell us if life is a cosmic imperative, happening any-where that conditions are favorable, or if it is a most improbable accident

After being launched on December 4, 1996, NASA's Mars Pathfinder impacted the surface on July 4, 1998, and bounced about 50 feet (15 meters) into the air, bouncing another 15 times and rolling before coming to rest approximately 2.5 minutes after impact and about 1 km from the initial impact site. The Mars Pathfinder's Sojourner Rover rolled onto Mars's surface on July 6. The Twin Peaks are modest-size hills to the southwest of the landing site. They were discovered on the first panoramas taken by the camera and are approximately 100 feet (30 meters) tall and less than a mile away. NASA selected the ancient flood plain on Mars as the landing site for the variety of rock and soil samples it may present. Eons ago, when water flowed on Mars, great floods inundated the landing site, located on a rocky plain in an area known today as Ares Vallis. (NASA/JPL/Caltech)

not to happen again on a billion planets in a billion years. A fundamental question indeed.

In contrast to Mars, Venus, our "sister" planet, is similar in size and mass to the Earth, but this is where all similarity ends. Venus developed a dense atmosphere also composed mostly of carbon dioxide with some nitrogen that eternally enshrouds the planet and does not allow a direct view of its surface. Clouds of sulfuric acid grace the Venusian sky (just in case you felt like flying there), and the atmospheric pressure on the surface is 100 times higher than Earth's. The greenhouse effect heats the surface way beyond what you would expect for its being 30 percent closer to the Sun than the Earth is, reaching temperatures high enough to melt lead. Venus does not look very promising as a place to search for life, at least not life as we know it. Between 1970 and 1985 ten spacecraft landed on the barren surface of Venus (all were Soviet spacecraft). Venera 9 was the first of these, surviving for just less than one hour on October 22, 1975, and sending us the first photo from the surface of this hellish place.

ВЕНЕРА-9   22.10.1975   ОБРАБОТКА ИППИ АН СССР   28.2.1976

The Soviet Venera 9 was the first robot to land on the surface of Venus. It survived the harsh conditions for just under one hour on October 22, 1975, and sent us a photo, a panoramic view from the surface of this hellish place. It was also the first-ever photo from the surface of another planet. The white disk at the bottom is part of the spacecraft which landed on the rocky surface. The horizon is visible on the top edges of this view. (NASA)

Frozen worlds

Giant Jupiter orbits the Sun at a great distance – five times farther than the Earth – so large that our Galileo spacecraft needed six years to reach this world, far beyond the habitable zone of the Sun. Surrounded by four large satellites, Jupiter looks like a miniature solar system. Its "planets" are frozen worlds, and to our great surprise recent explorations by Galileo tell us that they might not be frozen solid. Might Jupiter have its own habitable zone? The satellites were discovered by Galileo (the man), who in 1610 wrote in *Sidereus Nuncius*:

> Truly great are the things that in this short treatise I propose for the vision and contemplation by the students of nature. Great, I say, be it for the excellence of the subject matter itself, be it for its novelty never heard of before in all past time, be it also for the instrument, by virtue of which these things have become revealed to our senses . . . on January 7 of the present year 1610, on the first hour of the following night, while watching the celestial stars with the looking glass, Jupiter presented itself to me and since I had prepared for myself a particularly excellent instrument, I realized (that is, I had not realized this before because of the weakness of my previous instrument) that there were three truly small but still very bright starlets to its side, which although I believed them to belong to the fixed stars, still aroused certain marvel because of the fact that they appeared along a straight line parallel to the ecliptic . . . having returned, led I do not know by what fate, to the same investigation on January 8, I found a very different arrangement: all three starlets were in fact to the west of Jupiter and closer together than on the previous night . . . it was

The frozen worlds of Jupiter (from left to right: Io, Europa, Ganymede and Callisto) are as large as some of the terrestrial planets. Europa is slightly smaller than the Moon, whereas Ganymede is somewhat larger than Mercury. They are shown here as a montage with the correct relative sizes and the correct order from Jupiter, Io being the closest to Jupiter. They were named the "Medicean planets" by Galileo, who was trying to get a job in the court of the grand duke Cosimo II de Medici in Florence (he was successful). The names were suggested by Simon Marius, who observed the moons at the same time as Galileo, and were officially adopted in the mid-1800s. They orbit Jupiter quite rapidly, with periods that go from 1.8 days for inner Io, to 16.7 days for Callisto. You can see them with binoculars and observe how they move from one night to the next. (NASA/JPL/Caltech)

therefore established by me beyond any doubt that there are three stars in the sky wandering about Jupiter as do Venus and Mercury about the Sun, finally observed as clear as daylight in many other observations which followed: not only three but four are the wandering stars that complete their orbits about Jupiter . . .[3]

The Galilean moons were also observed by Simon Marius (1573–1624), a German contemporary of Galileo, who first proposed naming them Callisto, Ganymede, Europa and Io. Modern exploration of these satellites of Jupiter began with the launches of the Voyager spacecraft in 1977. NASA's Galileo mission, which arrived at Jupiter in December of 1995, has continued this exploration providing us with astonishing pictures of these fascinating worlds. Each of these worlds is unique. Callisto, the outermost one, has a frozen dark surface saturated with the scars of many impacts. Ganymede is farther in. It is the largest moon of the solar system, larger than the planet Mercury. Europa is next, somewhat smaller than our moon, with an icy, smooth, young surface marred by very long cracks in the ice and few craters. Io is the innermost one, slightly larger than Earth's moon. It is geologically

---

[3] *Sidereus Nuncius* by Galileo Galilei, a cura di Andrea Battistini, Marsilo Editori, 1993, translated by the author.

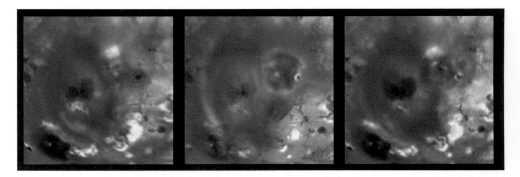

These three images of the volcanically active Pillan Patera region of Io show dramatic changes over a three-year period. The left image was obtained in April 1997 by the Galileo spacecraft, showing a red ring of material roughly 1000 miles across ejected by a volcano named Pele Patera, probably containing some form of sulfur. The middle image shows the same area in September 1997 after a huge eruption occurred. The eruption produced the large, dark deposit just above and to the right of the center. The deposit, which is 250 miles (400 km) in diameter, surrounds the volcano Pillan Patera and covers part of the bright red ring, which is the deposit from Pele's plume. The image on the right obtained on July 1999 shows the same area, but now the red material from Pele has started to cover, but has not yet entirely obscured, the dark material around Pillan. This indicates that both Pillan and Pele are still active. Each of the images covers an area approximately 1000 miles (1600 km) square and were taken from a distance of roughly 300 000 miles (500 000 km). (NASA/JPL/Caltech)

active with volcanoes which spew out 30-mile-high plumes of sulfurous compounds into space. The surface of Io, quite different from that of any other object in the solar system, looks to me rather like a cosmic pizza. Volcanic activity on Io tells us that, although far from the Sun, it must have a hot interior. Because of the large and changing tides produced by the gravitational pull of Jupiter, as Io orbits every 1.8 days, its interior is heated, in the same way that, if you twist a wire back and forth repeatedly, it will become hot.

Detailed images of Europa's frozen surface suggest that it resembles pack-ice found on the polar seas of our planet. The cracks, often interrupted by other cracks, and the lack of craters tell us that we are looking at a geologically young surface. The experience with Io suggests that Europa is also heated internally by tidal effects, in which case it could well be that, under some miles of surface ice, liquid oceans could be found. This would explain how pieces of its icy surface could have moved and why the few impact craters on its surface are unusually shallow. Jupiter, which even today radiates some internal heat, was much hotter 4 billion years ago, possibly sufficiently so to maintain a habitable zone at the orbit of Europa. Life could have started there, but as Jupiter cooled and Europa froze, conditions became less friendly. However, the extremophiles on Earth have taught us that life can flourish in what for us would be unfriendly conditions, and so the thought that some form of life inhabits the oceans under Europa's frozen surface is not as far fetched as it first sounds. The question then naturally arises: could it be that life developed on Europa and is currently to be found in its possible oceans?

It would be fantastic in both meanings of the word: unusual and unreal. No matter how life got started, no matter if it uses our biochemistry or not, we would find strange marine creatures that would be bizarre in many respects, maybe more exotic than the remains of the life forms of another world we *have* found: the fossils of the Burgess Shale. But we can also expect features that will be common to life evolved anywhere, because they are determined by the laws of physics and chemistry, universally valid. For example, we can expect underwater creatures to have shapes similar to the ones we are familiar with, because these are determined by the laws of hydrodynamics and by the clear advantage, valid everywhere, secured by moving with ease and a minimum of energy. We can expect animals with no eyes, since eyes are useless in the eternal darkness of the deep oceans unless they evolved some form of bioluminescence as we see in some terrestrial animals. Underwater sound is the way to communicate, as the whales do, and this is also a means of finding your way using sound echos. We will know what secrets lie beneath the surface of Europa by continuing exploration of this Galilean satellite. A positive answer to the question of life there would fittingly complete the Copernican revolution which began with the observations by Galileo of these moons of Jupiter. They would now be telling us that we are also not alone. We might have found a similar environment much closer to home: Lake Vostok.

Lake Vostok, one of the largest lakes on Earth, is 125 miles long, 40 miles wide, and 1500 feet deep, about the size of Lake Ontario. No one that we know has ever navigated its waters, in fact, no one has ever directly seen this large lake. This is because it is covered by more than 2 miles of Antarctic ice and has therefore been isolated from the biosphere for millions of years, perhaps as many as 40 million. The first clue of its existence came from aircraft radar and seismic studies done in the 1970s, but it was not until the 1990s that measurements from satellites confirmed that the lake was there.

Lake Vostok is no ordinary lake. The pressure on its surface is 350 times higher than the atmospheric pressure because of the 2 miles of ice piled on top of it. At this high pressure the water contains almost no dissolved oxygen or any other gas, so we do not expect to find life similar to that in other lakes. As we have seen, however, life will flourish in unlikely places such as the

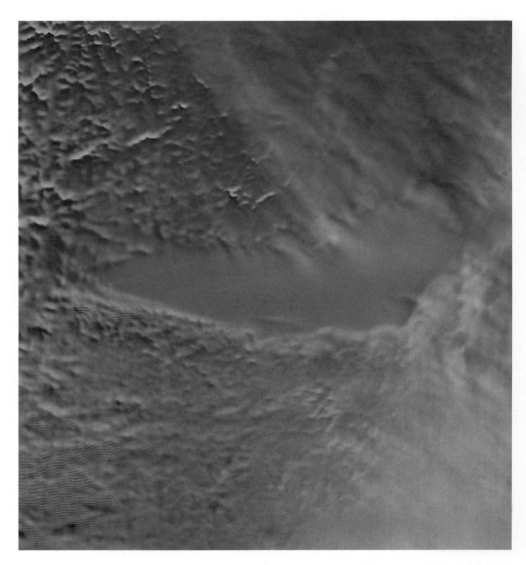

The outline of Lake Vostok, hidden beneath more than 2 miles of ice, is clearly seen in this image obtained by the Canadian RADARSAT satellite. This satellite uses radar to map features in the ice with high resolution. The very thin white line crossing the surface above the lake is a road across the ice leading to the Russian Antarctic Vostok research station at the left. The white dash just visible is the station's airport runway. (NASA GSFC Scientific Visualization Studio)

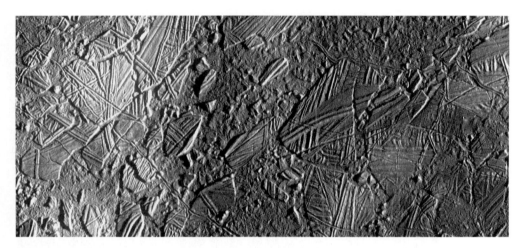

This image shows a small region (approximately 44 by 19 miles (70 by 30 km)) of the (disrupted) ice crust in the Conamara region of Jupiter's moon Europa. The white and blue colors outline areas that have been blanketed by a fine dust of ice particles ejected at the time of formation of a large (16 miles in diameter) crater, named Pwyll, some 600 miles (1000 km) to the south. A few small craters can be seen associated with these regions. These were probably formed, at the same time as the blanketing occurred, by large blocks of ice thrown up in the impact explosion that formed Pwyll. The unblanketed surface has a reddish brown color that has been painted by mineral contaminants carried and spread by water vapor released from below the crust when it was disrupted. (NASA/JPL/Caltech)

hydrothermal vents deep in the oceans, so it would surprise no one to find some kind of life somewhere in the lake. The water is kept liquid by the heat from inside the Earth, isolated by its thick blanket of ice. Scientists studying the evolution of life on Earth have been greatly interested in studying Lake Vostok. Because the lake has been cut off from the rest of the Earth for such a long time, any life in the lake will have evolved in peculiar and interesting ways. Scientists from the Russian Antarctic research station Vostok have drilled to a depth of 11 880 feet, about 300 feet from the water, stopping there to avoid contamination of the lake, until strict procedures to avoid this, a very difficult task, are carefully worked out. We also worry about con-taminating Europa, to the point that the trusty Galileo spacecraft will be commanded to plunge into Jupiter, once its fruitful mission is over, to avoid the slightest possibility that it could collide with one of the Galilean moons. Already, in deep ice cores drilled over the lake and dated to be 400 000 years old, bacteria that we have never seen before have been found, and when the drills eventually open this lost world it might be full of surprises.

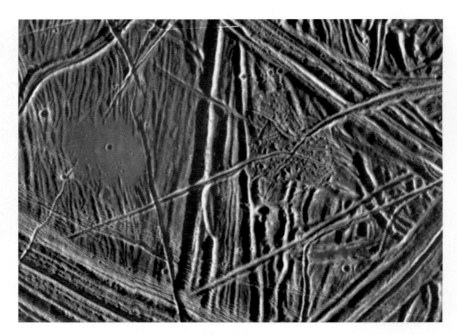

This close-up view of the icy surface of Europa, covers an area of about 7 by 10 miles (11 by 16 km) and has a resolution of 28 yards (26 meters). It was obtained in December 1996, by the Solid State Imaging system on board the Galileo spacecraft. A flat smooth round area about 2 miles (3.2 km) across is seen in the left part of the picture. It is thought that this area resulted from flooding by water which erupted onto the surface and buried sets of ridges and grooves. The smooth area contrasts with the rugged patch of terrain farther east, to the right of the prominent ridge system running down the middle of the picture. Eruptions of material onto the surface, crustal disruption, and the formation of complex networks of ridges show that significant energy is available in the interior of Europa. (NASA/JPL/Caltech)

Exploration of Lake Vostok might give us a taste of things to come in Europa and serves as a rehearsal for the future missions to this frozen world.

The second largest satellite in the solar system is Titan, Saturn's largest moon, slightly larger than Mercury. It is the only moon in the solar system to have a significant atmosphere which hides its surface. This atmosphere is mainly composed of molecular nitrogen (N_2) with a small percentage of methane (CH_4) and traces of hydrocarbons such as ethane (C_2H_6) and acetylene (C_2H_2). Although it has less than one half the mass of Mercury, Titan can retain this atmosphere of heavy gases because it is so much colder there, being ten times farther from the Sun than the Earth is. Lakes of liquid ethane and methane are believed to be on the surface, and if we ever arrive there we must take along "no smoking" signs, in case there is some oxygen

Majestic Jupiter's swirling atmosphere looms in the background as Io moves in its orbit illuminated by the Sun from the left. The Cassini spacecraft captured this image at the dawn of the new millennium on January 1, 2001. Cassini was launched in October of 1997 and will reach Saturn late in 2004. If all goes well, the Huygens probe will descend via parachute onto Titan's surface to return data and images that will lift the veil surrounding this exotic and distant world. (NASA/JPL/University of Arizona)

present. A probe called Huygens, launched from a spacecraft called Cassini, will descend into Titan's thick atmosphere when it arrives there on July of 2004 following a seven-year voyage. It will study the detailed properties of the atmosphere during its three-hour descent by parachute to the surface of this exotic world. Christiaan Huygens (1629–1695), a Dutch physicist and astronomer, discovered both Titan and the rings of Saturn in 1655. Jean Dominique Cassini (1625–1712), first director of the Paris Observatory (born in Italy as Giovanni Domenico), studied the rings of Saturn, discovering their gaps, and also discovered four of its moons. Exploration of Titan offers us the possibility of traveling back in time to encounter conditions which in many respects resemble those on Earth 4 billion years ago. Stay tuned then: we may have interesting news from Europa and Titan in the not too distant future.

Beyond the solar system

Without doubt, the Earth is the only planet with intelligent life orbiting our Sun. However, as we have seen, the Sun is quite an ordinary star, similar to approximately 10 percent of all stars. Although, as we saw in Chapter 3, we do understand in general terms the process by which stars and planets are formed, there was until recently no evidence that planets really had formed about stars other than the Sun. This has changed dramatically with the discovery, in the last few years, of disks of gas and dust around stars, the precursors to the formation of planets and, more significantly, evidence of

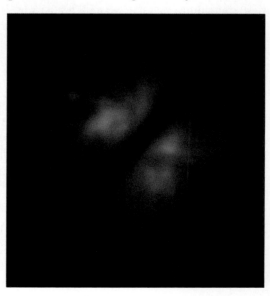

NASA's Hubble Space Telescope's Near-Infrared Camera and Multi-Object Spectrometer (NICMOS) took this image of a newly formed star, named IRAS 04302+2247, 450 light-years away in the constellation Taurus. The star is hidden from direct view and seen only by the reflected light from the nebula it illuminates. A dense, edge-on disk of dust and gas which appears as the thick dark band crossing the center of the image, divides it in two. The disk has a diameter 15 times the diameter of Neptune's orbit, and has a mass comparable to the Solar nebula, which gave birth to our planetary system. (D. Padgett (IPAC/Caltech), W. Brandner (IPAC), K. Stapelfeldt (JPL) and NASA)

planets orbiting about a growing number of stars. Current telescopes cannot easily directly detect a planet orbiting about another star because planets are billions of times fainter than stars and so are lost in a star's bright light. However, instruments will soon become available to do this, that is, if science keeps getting enough money from the appropriate government agencies.

We have indirect evidence that planets orbit other stars from measurements of the tiny gravitational effects which they exert on their stars. As an example, Jupiter, the most massive planet of the solar system, makes our Sun wobble in space as Jupiter circles it in its 11.8-year-long orbit. This is a consequence of the fact that, as Jupiter orbits the Sun, the Sun orbits Jupiter. In other words, both move about a point called their center of mass. For two objects of equal mass, the center of mass is located at the midpoint of the line joining them. Since the Sun has 1000 times the mass of Jupiter, their center of mass is 1000 times closer to the Sun than to Jupiter and the Sun moves in a circle 1000 times smaller than Jupiter's orbit. Thus, the size of this circle is roughly the same size as the Sun, meaning that the Sun just oscillates around its center. With current technology, we cannot detect this wobble for other stars directly – they are too far away – but astronomers have measured the tiny changes in the speed of a star as it sways back and forth relative to Earth.

From these measurements, we can derive the masses of these extra-solar planets and determine their orbits. However, we cannot observe them directly and therefore do not know anything about their particular properties. Since more massive planets are easier to detect because they exert a bigger effect on their star, the planets so far detected are massive, most of them more massive than Jupiter.

For one star, called HD209458 (the 209458th star in the Henry Draper catalog), a star similar to the Sun at a distance of 153 light-years, direct evidence for a planet has recently been obtained. This was possible because the planet orbiting this star happens to pass in front of it from our point of view. It therefore became feasible to observe the very slight decrease in the star's brightness as the planet crossed in front of it, an exciting discovery which removes any doubt about the interpretation of the observed wobbles as due to planets. So, whereas ten years ago nobody could say for certain that there were planets around stars other than the Sun, today the list of stars that we know to have planets is steadily growing. It is likely that millions of Earth-like planets orbit the stars of the Milky Way. We shall need new instruments to answer the questions that immediately spring to mind: do some have atmospheres containing oxygen, methane, and ozone? Do they have oceans of water on their surfaces? If so, are they the abodes for extraterrestrial life? And might this life be intelligent? Exciting questions whose answers I would like to know tomorrow.

SETI

As we have seen, the development of intelligence on Earth hinges on a complex chain of evolutionary contingencies and events, both astronomical and geophysical, which have made our planet what it is today. Take away our Moon, change the Earth's orbit slightly, let the continents evolve differently and we get a very different planet, maybe populated by just bacteria. Humans are the only one of billions of species which have populated our planet – most of them extinct – to have developed a certain degree of intelligence and a complex technology. Most living things live quite successfully without this, and this might be the case for the entire galaxy. On the other hand, the thought that Earth is the only planet with intelligent life in the entire galaxy does not seem to make sense. We shall know better through further research.

It can be argued that intelligence provides individuals with an advantage for survival, and hence will develop in the course of evolution on any planet. However, it could be that, just as a fire will extinguish itself after a time, a species which develops a level of intelligence where it has power over every other species and the environment will extinguish itself in due course. Once a life form is sufficiently intelligent to dominate and conquer the biosphere – and thereby destroy it – evolution will stop. Unfortunately, this intelligence will not have been sufficient to allow the life form to understand that it should not do this – until it is too late. The only example we have of this, namely us, hints at the truth of this proposition. Intelligent life might survive if there is some way to jump over this *"not-sufficiently-intelligent"* hurdle and reach a higher level.

So maybe, just maybe, we are not alone. Let us then assume that somewhere orbiting about a star similar to the Sun, hundreds or even thousands of light-years away, another planet similar to the Earth developed intelligent life. It is beyond our capacity to describe what these beings would look like, and this is in any case only important for Hollywood productions. From what we know, such beings would most likely be carbon-based life forms, with some type of similar biochemistry to ours, given that nature seems to be so predisposed. Some people might say that even this is not necessary, that this is just our parochial view of things and that beings could be silicon-based instead. Although silicon is not as chemically versatile as carbon, this might be so. (Perhaps there is a future on Earth occupied by silicon life, intelligent machines, descended from us but with sufficient intelligence to survive, with all the attributes of life. We will know how they are when *they* tell us.)

There are a few things, apart from death, about which we can be certain, or at least as certain as we are about other things we are certain of. We are certain that the laws of nature that we have discovered are truly universal, valid at all places and at all times, at least all those we are concerned with here. If this were not so we would not be able to understand anything. Gravity acts the same way around any star anywhere, like charges repel each other wherever they might be, the chemical elements are the same and electromagnetic waves, of which light is just a small fraction, behave the same everywhere. Heaven and Earth *are* the same. We also claim that 2 plus 2 is 4, as much on Alpha Centauri as here. If this were not so we would not be able to understand anything.

An intelligent life form will presumably develop a technology based on these natural laws, which would inevitably lead to tools similar to those that we have built. I do not mean specific items such as cars and refrigerators, but the general aspects of technology with methods to harness and produce energy and the development of systems for transportation and communication. Communication is a fundamental feature of intelligent societies, at the basis of our social life and development. We began with cave drawings and clay tablets, then Johannes Gutenberg (about 1390–1468) invented the printing press, and early last century we developed the technology to communicate over long distances using radio waves. An ever growing web of cables, optical fibers, and satellite links distribute information all over the globe, as if they were our planet's nervous system. We are striving to reach the point where information becomes accessible to all people. It is astonishing how today we are able to look into the menu of a restaurant in Paris from somewhere halfway around the world. It is reasonable to suppose that this development, reflecting a natural characteristic of intelligence, may have taken place elsewhere. If it has, then perhaps we might be able to "eavesdrop" with our telescopes, especially those sensitive to radio waves, such as the giant one at Arecibo, and find some indication of their communications. This is what the search for extraterrestrial intelligence (SETI) is all about.

One further thing we can say about possible intelligent extraterrestrial beings is that they will be much more advanced than we are. This is so because, as we have seen, in the five-hour movie about the story of our planet (from birth to death) we have appeared for one second at the end of three hours of history. Finding others in the Galaxy who are also precisely at this first second of their journey would be highly unlikely.

One of the first things I would ask an extraterrestrial would be: how did you manage to survive, that is, how did you pass the "not-sufficient-intelligence" hurdle? The answer might be the most important answer ever. After this there would be 1000 other important questions just to break the ice. What would you ask? Unfortunately, even if these extraterrestrials are from a world oribiting a star quite near to the Sun, the answer will take many years to arrive, by which time we might have forgotten what the question was, never mind the problem of languages. If it were a distant star there might be no one left to record the answer. So the idea that we could hold a "conversation" is quite unreal.

It might also be reasonable to suppose that after a few seconds of their cosmic history an intelligent life form ceases to use radio waves to communicate, ceases to use high-power broadcasting, and develops a system where all communications are efficiently routed via something like optical fibers, so that very little energy is wasted and nothing is "lost" into space. The point is that advanced civilizations might not accidentally leak radiation into space, or will at least reduce this, as we are starting to do on Earth. Thus, advanced extraterrestrial societies, even if they are numerous, might not be as easy to detect as we may suppose, unless they wish to be detected for whatever reason and deliberately set up a beacon.

We did this once in 1974, when on November 16, to celebrate the installation of a new reflector, a brief message was sent from the Arecibo telescope. The 3-minute-short message was beamed toward the globular cluster M13, one of a couple of hundred such objects that surround our galaxy. This cluster, a giant spherical conglomerate of half a million stars, is at a distance of about 25 000 light-years. If someone, 25 000 years from now (we won't quibble about the 30 years already elapsed), on a possible planet in orbit about one of the stars in M13, happens to point a radio telescope in our direction just when our message reaches them (it has 3 minutes within which it can do this) then we might get an answer 50 000 years from now, give or take a few thousand. Believe it or not, some people were upset about this symbolic act, claiming that we were giving away our position to potential cosmic predators. For me, this is a reflection from the mirror we look into, more than a reflection of reality. We need to worry about things much closer to home, as we shall see in the next chapter. In any case, for more than fifty years we have been revealing our existence to anyone out there, as a sphere filled with waves carrying our radio and TV programs expands from Earth at the speed of light. Anyone detecting these could conclude that there is a primitive technology on Earth. Whether, upon deciphering the content of

The globular cluster M13, or NGC6205, in the constellation Hercules, is one of the most noticeable globular clusters in the northern hemisphere. It is about 25 000 light-years away and about 150 light-years across, containing perhaps half a million stars. The stars in this cluster are over 12 billion years old. About 25 000 years from now the most feeble wave will indicate to anyone able to perceive it, on some planet of one of the stars in the cluster, that they are not alone, that in the direction of a small yellow star that we call Sun, there was a technological civilization. (N.A.Sharp, REU program/AURA/NOAO/NSF)

the transmissions, they will also conclude that there is intelligent life on Earth is a different matter.

Recall also that a star with even only three times the mass of the Sun has a lifetime of just a few hundred million years, much shorter than the Sun, and a luminosity tens of times higher. So life had better hurry-up developing on a suitable planet in orbit about such a star, if it is to evolve by chance into something similar (or hopefully better) than us before the show is all over. Even worse, stars more massive than our Sun produce a much greater quantity of ultraviolet radiation, so ozone is not going to help very much, and life had better evolve some other means of protection, or stay deep in the oceans. But then it is hard to see how these creatures are going to develop electronics to communicate via radio waves, which in any case do not propagate great distances in the oceans, without thinking about such other details as problems with corrosion. This is in the realm of speculation, but it points out the kind of considerations that go into trying to understand all the factors which might be relevant to the discussion about the prevalence and detection of intelligent extraterrestrial life.

We will maximize our chances of detecting ET by searching for signals from planets orbiting nearby stars similar to our Sun, where by "nearby" we mean distances smaller than 100 light-years. This is what current SETI programs do. The first SETI project began in 1960 when Frank Drake (1930–), then a young radio astronomer, used a small radio telescope in West Virginia to search for extraterrestrial transmissions from two nearby stars: Tau Ceti and Epsilon Eridani. Since that pioneering experiment, many other searches have been made with increasingly greater sensitivity and more sophisticated instruments. It is interesting that a Jupiter-mass planet in orbit about Epsilon Eridani, a mere 10 light-years away, was discovered in 1988. Our efforts have only just begun to sample the thousands of stars and millions of radio channels where signals might be found. *To date, in spite of what many believe, none has uncovered a signal from another civilization.* As I mentioned before, this is not a denial to keep things secret, it is the simple truth. You should realize that keeping such a secret would not be possible, even if this were desired, since astronomical observatories are very open places.

One uncertainty of any SETI project, besides those I have discussed, is choosing which of the millions of possible frequencies to search. While most radio wavelengths can pass unperturbed through the gas and dust that permeate interstellar space, the centimeter wavelength region of the spectrum is considered the most likely to be used for interstellar communication. Naturally produced background radiation is at a minimum at these wavelengths, so that any weak signal will be easiest to detect. We saw that hydrogen atoms in interstellar space emit radio waves with a wavelength of 21 centimeters. The OH molecule, quite abundant in molecular clouds, is detected by radio telescopes at a wavelength of 18 centimeters. Water, the most important compound for life, is the combination of these two, and so just as the different animals of the African savannah meet at a water hole, it would be fitting if different civilizations of our Galaxy met at a wavelength somewhere between 18 and 21 centimeters, a "water hole" in the electromagnetic spectrum. This sounds metaphorical enough, but we have no way of knowing if the extraterrestrials might feel the same way.

If some day in the future the giant reflector at Arecibo should intercept radio waves originating from some unique direction in the sky that seem to be artificially generated, the priority will be to confirm that the signal is truly extraterrestrial, and not due to satellite signals, radar, or any of the steadily increasing gadgets we use for communications. Then these observations must be confirmed by other large radio telescopes around the world as they point in this direction in the sky to see if they detect the same signal. Arecibo will then announce to the world that we might have the first evidence of extraterrestrial technology. *It will be a day when Giordano Bruno's soul will smile.*

On this photo taken by the Apollo 11 crew in July of 1969 as humans reached and landed on the Moon for the first time, you can see the Earth as it rises over the Moon's horizon. We had come a long way! This photo, more than anything else, made us aware of the fact that we live in a limited world, passengers on a lonely planet, "Spaceship Earth," the only place we've got. (NASA)

Chapter 8

The dark crystal ball

A wonderful sense of doom and futility[1]

After a long evolutionary march filled with surprises and mishaps we find ourselves only a short distance away from a precipice. If we do not take our heads out of the sand and act energetically and with the necessary sense of urgency, our future might not be. Still, we keep blindly marching on, pushed by those behind us who cannot clearly see what is ahead. Only a bit more to go, until it is too late.

Warning

A statement prepared by the Union of Concerned Scientists, in 1992, and signed by prominent scientists from around the world, entitled "World Scientists' Warning to Humanity" said in part (the full text is reprinted in Appendix C):

> Human beings and the natural world are on a collision course. Human activities inflict harsh and often irreversible damage on the environment and on critical resources. If not checked, many of our current practices put at serious risk the future that we wish for human society and the plant and animal kingdoms, and may so alter the living world that it will be unable to sustain life in the manner that we know. Fundamental changes are urgent if we are to avoid the collision our present course will bring about.

Sadly, this warning is still true ten years later and will most likely be even more pertinent in the year 2050. Such is the slowness of the cure by any remedial action we might take.

Just 200 years ago we did not have a clue, not the faintest idea, about the nature of a star, or the nucleus of a cell. That there is a connection between

[1] Douglas Adams, *The Restaurant at the End of the Universe*, Six Stories, p. 227.

the two would have been considered sheer lunacy. Today, as we have seen, we understand the relationship between stars and life in great detail, and can be astonished at what we have learned. It is a wonderful story, the result of a great leap in our understanding of nature which occurred with increasing speed over the last few hundred years. This great achievement of the human mind has gone hand-in-hand with our short-term power over nature, which, as we shall see, is threatening the biosphere and our very existence. Not such a wonderful story. Although you hear a great deal these days about our global environmental problems, we should look at these now, under the new light provided by the cosmic perspective of the previous chapters.

Life on Earth has adapted to changes in atmospheric composition and surface temperatures and has survived tremendous environmental upheavals. However, the fossil record bears the remains of myriads of species which did not survive these trials caused by natural events and shows that most species that ever lived are now extinct. As we have seen, our species arose as a result of a mass extinction 65 million years ago, which wiped out the dinosaurs and about 75 percent of the species then living, but sparing some small mammals. In turn, life has affected our planet on a global scale, in particular changing the composition of our atmosphere, through billions of years of photosynthesis. Over these long time scales, life had ample time to adapt to these slowly changing conditions by the process of natural selection. When conditions changed rapidly, mass extinctions occurred. Today we are very much concerned about global changes to the environment caused by human activities. We are causing another mass extinction which affects all life forms on Earth. The difference this time is that *we* are the agents of destruction, the cause of this sixth mass extinction, we who claim a high moral standard and the highest status in the living world.

I imagine that eons from now the stratigraphic record will document this mass extinction. Someone, although I can't imagine who, will be looking at an ancient cliff from a distance and note that a strange thick and twisted mile-long layer of material of a different color paints its walls. Careful examination by a large team of scientists will slowly reveal the fascinating origin of these deposits. These are the amalgamated and metamorphosed remains of entire cities, mostly stone and steel with a few bits of very interesting materials of puzzling origin, squashed together by tectonic processes as if they were the body of a car which has been reprocessed by one of those special machines found in a junkyard.

Be fruitful and multiply

The human population of the Earth has dramatically increased from less than 500 million in the year 1000 to 6000 million in the year 2000. It was only 2000 million in 1930. This unprecedented *population explosion* is a major cause of many of our problems. The human population is expected to pass 9000 million by the year 2050, so our children will have to deal with a very crowded world. It does not matter where the increase is currently most rapid; it still affects the entire planet, although most resources are in fact consumed by a minority in the developed countries.

Every hour the population of the world *increases* by about 10 000 persons: 85 million every year. Just think about this for a moment; by the time you finish reading this sentence there will be another 30 humans added to the world population. This is the greatest problem our planet faces, much more urgent than the threat of a collision with an asteroid. In 1798, Thomas Robert Malthus (1766–1834), an English economist, articulated the idea that the growth of population always outpaces the growth of production and therefore, even from just a standpoint of available food, things would not turn out well. He could not have imagined the current environmental predicament in which we find ourselves. As an interesting aside you will realize, although probably you do not like the idea, that most of those 9 billion, living in 2050, will be others than the current 6 billion. Such is life.

Because of this rapid population increase, our technological devolvement which has given us tremendous power over nature, and our past lack of foresight, the biosphere has been subjected to colossal abuse. I am not writing this as an indictment of either our technology, or the social development which brought us this far. It would not have been possible for a Henry Ford to understand the ultimate consequences of the mass production of automobiles, any more than Columbus could have foreseen modern America. Thanks to some of our technological tools, such as powerful computers and remote-sensing spacecraft, and the great advances in our understanding of nature, we have today a much better grasp of the past and can forecast with a good deal of confidence what will happen in the future. We have recklessly introduced large quantities of pollutants into the land we live on, into the water so essential to life, and into the air that we breathe. You know it is so and, as I will explain later, the frogs are feeling it. Our activities, the effects of which rival the natural traumatic events of the past, are affecting large areas of our planet. In short, *we have become an environmental hazard*. The degradation of the environment even affects our ability to study the Universe inasmuch as we are creating a "fog" of light and radio waves

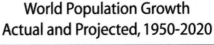

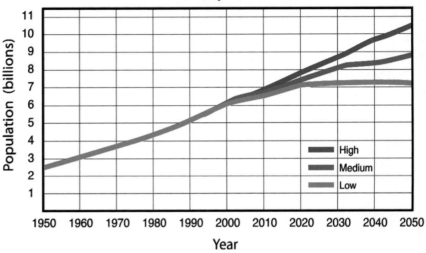

This innocent looking graph is probably the most scary diagram you could possibly show to anyone. It documents how our numbers have increased, and will increase in the future. We were only 3 billion in 1960 and had doubled our numbers by 1999. By 2050 we will have reached 9 billion, one-and-a-half times today's population, although depending on such things as the fertility and mortality rates it could be higher or lower as indicated. So, within 50 years, just to remain in the sorry state we find ourselves in now, we would have to increase by an equivalent amount our hospitals, schools, housing, transportation systems, water supplies, food supplies and all other goods and services which form the basis of our societies: all this just to make no progress at all. In addition the increase in population and associated goods would tax beyond limits the capacity of the global ecosystem to absorb the additional waste we will produce and the abuse we would subject it to. This growth in our numbers and its consequences represent the most dramatic change in the biosphere over all of our history. (José F. Salgado, after a graph of the United Nations)

that is increasingly impairing our vision. We could easily be indicted for the consequences of our actions, by anyone left to do so 100 years or so from now. However, even if convicted, we shall not be there to be punished, so the question is one of our moral sense of responsibility.

If we do not voluntarily control population growth and our inefficient use of valuable resources, nature will do it for us in the most painful of ways – the process has already begun. A large fraction of the population of the world is malnourished and starvation is a recurring theme, killing 1000 people every hour, directly from hunger, or indirectly by falling easy prey to disease. About 800 million people go to bed hungry every night and over one-quarter of Earth's population do not have access to dependable fresh water.

The 1970 Nobel Peace Prize was awarded to the American agronomist Norman Borlaug (1914–) for his leadership in producing the so-called Green Revolution, the result of research that led to new high-yield-crop varieties. This allowed world production of cereals to more than triple in the last 30 years. For some countries such as India and Pakistan, it resulted in a reprieve from massive starvation. But the success of the Green Revolution has been crippled by relentless population growth. As Borlaug stated in his acceptance speech three decades ago, when the Apollo astronauts made us clearly see "Spaceship Earth," and when the population of the Earth was only 3.7 billion:

> the Green Revolution has won a temporary success in man's war against hunger and deprivation; it has given man a breathing space. If fully implemented, the revolution can provide sufficient food for sustenance during the next three decades. But the frightening power of human reproduction must also be curbed; otherwise the success of the Green Revolution will be ephemeral only.[2]

How right he was!

Viruses and prokaryotes

Genetic engineering is providing plant varieties resistant to pests and diseases, and therefore contributes to increase yields. Still, genetically modified (GM) foods, already the subject of great controversy between those who see them as a way to feed the hungry and those who fear some unexpected ill-effects, will not solve our problems either. We are simply too many.

The increased population density and squalid living conditions in many areas of the world provide fertile ground for bacteria and viruses to prosper, while the increased mobility of humans favors transmission of diseases which in the past were easier to contain. Bacteria or viruses which infect a person in some distant land can arrive in your living room the next day, and this is not just in the movies, as recent events have shown. Viruses consist of genome (DNA or RNA) enclosed in protein without the cellular machinery for reproduction. This is why their status as living is ambiguous. They reproduce by invading a cell to use its reproductive machinery and in the process

---

[2] "The Green Revolution: Peace and Humanity," a speech by Norman Borlaug on the occasion of his award of the 1970 Nobel Peace Price in Oslo, Norway, on December 11, 1970.

often kill the cell. The newly created viruses go on to invade other cells. Left unchecked, and depending on the type of virus, this process can lead to dire consequences for the organism. This is the essence of a viral disease.

Ebola hemorrhagic fever is a deadly, terrifying, fast-acting virus. Symptoms of ebola begin a few days after infection, causing fever, sore throat, vomiting, diarrhea, abdominal and chest pains, severe bleeding, shock, and then after a couple of weeks, merciful death. All this is caused by a microscopic agent whose RNA uses our cells to replicate. The only good thing about ebola is that it is so virulent and fast-acting that a population can easily be identified and isolated to stop the spread of the disease. Recent outbreaks in Zaire and the Sudan killed several hundred persons and were then controlled. Ebola is transmitted by close contact with persons recognizably very ill with the disease. God help us from a mutation which takes months or even years before it acts, such as the Human Immunodeficiency Virus (HIV), which causes AIDS. This microscopic organism, a lentivirus, is composed of RNA and some enzymes surrounded by a protein envelope. Tens of millions of people worldwide have already been infected by this silent killer which is gaining ground every day, and is worst in Africa. It is estimated that one in four inhabitants of Botswana and Zimbabwe, with a total population of about 12 million, are affected. All of the countries in southern Africa face a devastating health crisis caused by this virus.

AIDS is particularly perverse because although deadly it is slow, and persons carrying the virus might not even know it for years, providing ample chance for infection. The disease started in Africa, where it is the largest cause of death, probably in the Congo by the transmission of a virus carried by a subspecies of the chimpanzee (*Pan troglodytes troglodytes*). This animal is in danger of extinction because thousands are killed every year to be eaten, which is probably how the virus made it to humans in the first place. Because the chimpanzee carries the virus, but does not develop AIDS, there is great hope that studies of these cousins of ours, which have a genetic makeup very close to ours, will provide the necessary information to develop a cure and a vaccine for this disease, that is, if we do not kill them all first.

Although AIDS has been at center stage for quite a while, there is no reason to think that it is a one-of-a-kind disease. New global epidemics are just waiting for the opportunity, and mutated strains of old diseases which show resistance to drugs and vaccines are waiting for a comeback. Although great advances in our fight against infectious diseases were made in the last 100 years, the recent outbreaks of hanta and ebola viruses, and antibiotic-resistant tuberculosis, threaten to undo our triumphs. The 1918–19 influenza pandemic killed some 25 million persons worldwide, comprising

1.4 percent of Earth's population which at that time was just 1.8 billion. If the same happened today it would kill 80 million persons.

The climate

Climate change has become a topic of great concern, as we have learned that it has been a major factor in the history of life on Earth, and will affect the future in ways that are difficult to predict. These days the weather has been uncommon; either it has been colder or warmer than it usually is, or it has been wetter or dryer, stormier or unusually calm. However, I am not talking about the weather here, but about global climactic changes which can affect life in a fundamental way. Of course, the climate changes naturally, as a consequence of several geophysical and astronomical phenomena which include the slowly changing configuration of the continents and small changes in the orbit of the Earth and its axis of rotation. We know of past ice ages, the last one ending about 15 000 years ago, followed by warm interglacial periods where the average temperature increased by about 5 °C (9 °F). This is not a large quantity, if you think about it: smaller than the usual changes we sometimes experience as we go from one day to the next. The difference lies in the fact that these are local – not global – changes.

By analyzing the composition of small air bubbles trapped in ice cores drilled in polar ice, formed from accumulated snowfall, we can measure the past concentration of greenhouse gases in the atmosphere. This provides a record to study past climate. At the Vostok Antarctic station, for example, ice from a depth of 3 300 m (10 800 feet) is over 400 000 years old. The recent record shows a steady increase in the concentration of greenhouse gases, starting about 1850, the beginning of the pre-industrial era, and it is understood to be anthropogenic (of human origin).

The great concern is that this is leading, by way of the greenhouse effect, to an increase in the average surface temperature of our planet. Carbon dioxide has increased by about one-third, methane by a factor of 2, and nitrous oxide by 15 percent. Although, as we have seen, carbon dioxide is only a minor component of the atmosphere, the total amount of carbon in the atmosphere is quite large: 750 gigatons of it. It is estimated that human activities currently add about 7 gigatons per year to the atmosphere, with about one-half of this remaining there, and that this will only increase in the future. Even if it does not, at this rate the concentration of carbon dioxide will double in about 200 years and this doubling is expected to increase the average surface temperature by 2–5 °C (4–9 °F). This sounds like a small quantity, and you might say that you would not mind a bit more

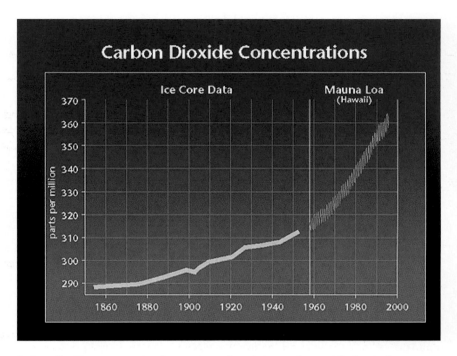

Carbon dioxide concentration in our atmosphere is increasing rapidly due to the burning of fossil fuels (coal, gas, and petroleum products). A historical record of carbon dioxide concentration can be obtained by analyzing bubbles of air trapped in polar ice. By drilling deep into the ice we can sample air from the past. The graph shows that in pre-industrial times the carbon dioxide concentration was fairly constant at about 280 parts per million (ppm) and began a steady and steep increase after about 1850, having increased by 30 percent. Results from both poles show similar trends, indicating that this is a truly global occurrence. (White House Initiative on Global Climate Change)

warmth, if you live in places like Boston or London, but this "small" change represents a 10 percent increase in the average surface temperature of the planet, with dramatic effects, not fully understood. The last ice age was the result of a lowering of the global average temperature comparable to the expected rise we are facing now.

Data from the Vostok ice core, and similar cores from Greenland, show a very clear relation between the concentration of carbon dioxide and methane on the one hand, and temperature on the other, as can be seen in the figure, where these quantities are plotted for the last 160 000 years. Note that the present CO_2 level is much higher than at any time in those 160 000 years. Past temperatures are obtained from studies of the ratio of the naturally occurring stable heavy isotope of oxygen (^{18}O) to the normal one (^{16}O) which is about 1 in 500. It happens that as water (H_2O) evaporates from the oceans, the water with normal oxygen, being very slightly lighter than that made with the heavier isotope, will evaporate more easily, and this depends very slightly on the ocean temperature. The remaining ocean water, or the water vapor in the atmosphere, will then contain more or less water with ^{18}O depending on the prevailing temperature. It is possible to measure the proportion of this isotope in the past since it is incorporated in the carbonates ($CaCO_3$) of the shells of microscopic sea creatures which formed ocean sediments and then eventually became rocks. These can be analyzed, and their ages determined with radiocarbon methods. The isotopic ratio in the water from ice cores can also be measured, providing another measurement of temperature. These are difficult measurements which require great care, but different studies, using samples in various locations, done by separate research groups, reach similar results, providing great confidence in them.

During the last ice age, the North American continent was covered by a 1-mile-thick sheet of ice which reached as far south as the Great Lakes. So a small change of even one degree can give rise to a large effect. There is uncertainty about how much the effect of doubling carbon dioxide will be and this is why the expected increase is said to be between 2 and 5 °C, and not a definite 3.5 °C. This is because computer models of such a complex system as the climate depend on a variety of quantities, some better known than others. A great variety of relevant physical and chemical processes must be well understood, including the complex cycles that link processes in the oceans, land, atmosphere, and the biosphere. Account must be taken of all possible feedback mechanisms. These can be *positive*, such that a change in one quantity affects another which in turn changes the first one in the same direction, therefore reinforcing the effect. Such positive feedbacks can lead to very rapid changes of great magnitude.

As an example consider the reflectivity of the Earth's surface, its albedo, which is one quantity that determines the average surface temperature. Ice and snow are very good reflectors of sunlight, so a larger ice cover will increase albedo, reflecting more sunlight back into space and cooling the Earth. If for some reason, such as a change in the direction of Earth's axis of rotation, the climate becomes colder and the area of snow and ice increases, this will reinforce the cooling and it will become still colder – positive feedback. If there were no other influences, this would continue until we had a completely frozen, and dead, Earth.

Fortunately, there are also *negative* feedbacks, that is, effects which counter the original change. In the example we are considering, as the Earth freezes there will be much less water evaporation and eventually rainfall will stop. Without rainfall, carbon dioxide, released by volcanoes, will not be

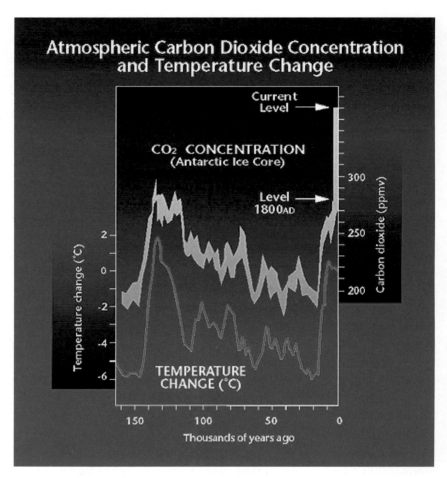

This graph shows some of the information which has been obtained from the Vostok ice core in Antarctica. The core has been drilled to a depth of 11 880 feet (3622 meters), just 300 feet (100 m) above the surface of lake Vostok. It provides data over a time span of 420 000 years. The most recent 160 000 years of data on atmospheric carbon dioxide concentration and average temperature are shown on this graph. We can see two striking features: first, the overall trends over a span of 160 000 years for each graph is maintained within bounds, for example CO_2 is never over 300 or under 180 ppm (parts per million by volume); second, we can clearly see that the variations are not independent, CO_2 concentration changes clearly correspond to changes in temperature, high CO_2 corresponding to high temperature. You can see two interglacial periods, a period of high temperatures that started quite rapidly about 140 000 years ago, and the current one which started about 15 000 years ago. Of great concern is the high value of the current CO_2 concentration (360 ppmv), and the predicted increase to 700 soon, very much higher that at any time in the last 160 000 years (in fact higher than over the entire 420 000 years of the ice core). (White House Initiative on Global Climate Change)

removed from the atmosphere and will steadily accumulate. This will lead to an increase in temperatures, through the greenhouse effect, which will in time reverse the global freeze.

It is remarkable that scientists have found evidence that this indeed happened to Earth prior to 550 million years ago, in the Precambrian. There is evidence for at least two such "snowball Earth" episodes, lasting over 10 million years each. The first one happened about 2.4 billion years ago and the second one between 800 and 600 million years ago, just before the Cambrian explosion. During this time Earth's oceans were covered by a thick layer of ice and global temperatures fell to below $-20\,^{\circ}C$ $(-4\,^{\circ}F)$. In this case, our hot radioactive core, courtesy of a long-past supernova explosion, caused, through volcanic emissions, an enormous increase in atmospheric carbon dioxide, which saved our planet from becoming a lifeless ball of ice. Only because the ice cover was thin enough at the equator for light to reach the ocean waters below were some photosynthetic algae and bacteria able to survive, so that life was able to recover from this global catastrophe.

Although there is a narrow range of outcomes in the computer programs used to model the climate, there is little uncertainty about the fact that it will get warmer, and more so, the more greenhouse gases we spew into the atmosphere. We shall have to face a changing climate caused by increasing temperature. This will lead to rising sea levels, mostly from the thermal expansion of the volume of water, and partly from the melting of glacial ice. By the year 2100, average sea levels will possibly have risen by 1 foot or more, the actual figure depending on how much the temperature rises. Since over one-third of the population of Earth resides in coastal areas it will be significantly affected by this increase in sea level. Precipitation patterns will also change, with not-yet-fully-understood consequences for an already precarious food production system, and a long and complicated chain of events could unfold creating significant disruption to societies all over the globe. Diseases such as malaria, dengue, yellow fever, and cholera thrive in warm climates, so the increased temperature will increase the occurrence of these afflictions. Although there are people, including scientists, who regard global warming as not yet proven, the evidence within the uncertainties points clearly in this direction. Millions of direct temperature measurements all over the globe have been analyzed by several research groups in different countries and the results coincide. The Earth has been getting warmer by a small but significant amount. The snows of Kilimanjaro *will* melt. Reducing the emission of greenhouse gases at any costs must be among the highest priorities of all governments in spite of the predictable objections by the industrial lobby. The cost of not doing so will be much higher.

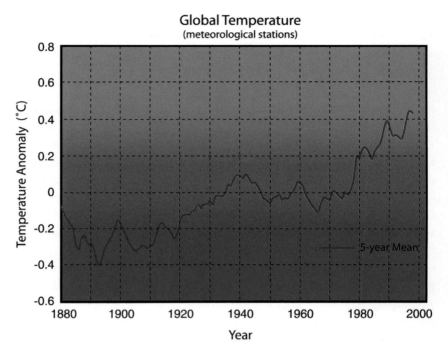

This graph shows the global annual average surface temperature, collected from several thousand meteorological stations around the world between 1880 and 2000, plotted as the difference (temperature anomaly) with the average for 1951 to 1980, which is the line at zero. You can appreciate that, although there are natural fluctuations from year to year, all the temperatures after 1980 were above this average, and all the temperatures before 1930 were below this average. The temperature has been slowly rising over the last 100 years, and this rise has been more rapid in the last two decades. The blue line is obtained as a five-year average and shows the trend clearly: global warming is not a matter of opinion. (José F. Salgado after a graph by NASA)

Biodiversity

About 6 percent of Earth's land area is currently covered by tropical rain forests that are the habitats of a great number of species, and this area is decreasing at the alarming rate of 1 percent per year. It is estimated that every year about 10 million hectares (about 40 000 square miles) of tropical forests are cleared, to provide wood or for cultivation. This corresponds to an area similar to that of the state of Virginia being cleared every year, a proposition as true as it is incredible. Or imagine this: on a pleasant winter afternoon, with sunny, clear skies, you find yourself on a hill looking out over a beautiful forested land. You can see green hills for miles in any direction. As the Sun heats the forest, the sweet perfume in the air makes you feel both good and a bit sleepy. Feeling so relaxed, you fall asleep in a hammock which just happens to be there in a nice sheltered corner. You wake a couple of hours later. As if in a nightmare, the beautiful forest surrounding you is gone, you see only bare land, and several miles away you can make out a wave of destruction clearing the land all the way to the horizon. The green hills are gone. In this very short parable I could say that it really was a nightmare while you were asleep. However, this nightmare is very real. Every two *seconds* a football field worth of tropical forest is cleared, destroying valuable plant and animal habitats, and reducing biodiversity. At the current rate of deforestation there will be very little virgin forest left in a couple of hundred years, endangering the stability achieved during a billion of years of photosynthesis, as forests supplied oxygen and removed carbon dioxide from the atmosphere allowing life to migrate to the land in the first place.

It is estimated that there are roughly 10 million species on Earth, but it could be much higher; we simply don't know. One ounce of forest soil contains thousands of species of bacteria, most of which we have not studied. We have catalogued less than 2 million species (about 750 000 are insects), and have studied only a small fraction of these in some detail. It is estimated that every day about 100 species of plants and animals become extinct, forever taking with them a valuable resource which has been at the origin of many important pharmaceuticals and new varieties of food. About one half of all medical prescriptions made in the United States are for drugs derived from natural products. We do not know if one of the organisms which will disappear tomorrow did not have in it a compound which is the ingredient for the cure of an "incurable" disease. It is our loss.

To succeed in the future, we need to learn as much as we can from the legendary books of the Sibyls, which contained supposed prophecies of divinely inspired seeresses. Unfortunately, every time that a species becomes extinct, it is like ripping out a random page from one of these books. We do not know how important it might have been for us to understand the contents of that page. If, for example, we lose the chimpanzee, of which fewer than 100 000 are left, we might lose our best chance for discovering a cure for AIDS.

However, it is not only the loss of potential products that should concern us. Biodiversity plays an important overall role in the biosphere because of a series of complex interconnected processes, which purify water, enrich the soil, process waste, and clean the air, supporting higher life forms and stabilizing our ecosystems. We are part of an intricate network of life forms, some living symbiotically with us, a relationship that had its origins when the cells forming our bodies started life as a cooperative venture between

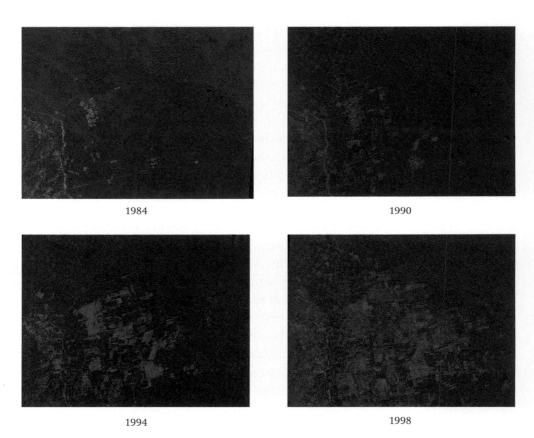

1984   1990

1994   1998

Since the invention of agriculture about one-third of the forest cover of our planet has been lost. Forests are being destroyed at a rapid pace leading to a great loss of biodiversity with consequences that we have not yet fully understood. If current trends continue, most of the tropical rain forest will be gone by the end of this century and different species of animals and plants will be irretrievably lost. About one-third of the world's tropical forests are in Brazil and surrounding countries. These images show deforestation from 1984 to 1998 near the growing city of Santa Cruz, Bolivia, which lies to the west of the river cutting through the scene. (North is up and west is left.) The scene is roughly 150 miles tall by 200 miles wide. In the initial 1984 scene, some clearing has already occurred in the humid forest. (NASA GSFC Scientific Visualization Studio)

some bacteria. You might think that we would be better off without some species of bacteria (such as tuberculosis and smallpox) or insects (such as the exasperating mosquito), but biodiversity is not a matter of saving one species out of millions. We do not know which species are crucial for the maintenance of other species and ultimately for us, and so the loss of biodiversity, with unknown consequences, entails a very high risk. We, who are at the top of the food chain, cannot survive without the services provided by the rest, whereas most of them don't need us, and, as I have suggested, might be better off without us. Beyond this, if we somehow manage to survive, it will be a much impoverished world to live in if the only remaining tigers, elephants, cheetahs, or whales are the sad inhabitants of a zoo.

Frogs are amphibians – animals which can live in water and in air – descendants of those that, 400 million years ago, in the Devonian period, ventured onto land. It was a bold move to exploit the favorable conditions in and around the water's edge. All over the world scientists are finding that the population of frogs is declining and species are disappearing. These animals, which live in a great variety of habitats, from the humid tropical forests to the Arctic circle, survived several mass extinctions, including the last one 65 million years ago. Today, many species of frogs are becoming extinct at an alarming rate, even in remote and pristine places where human influence on the environment seems minimal.

Frogs, like all amphibians, absorb chemicals and exchange gases directly through their moist skins and are therefore very sensitive natural sentinels of the environment. And what they are telling us, as they die, is disturbing. In the same way that smoke detectors give us early warning of fire, the disappearing frogs are sounding the alarm as they detect subtle poisons in the biosphere, which have yet to affect us.

Whereas, over time, repairing the physical damage we are inflicting on the biosphere might be possible, the loss of biodiversity is irreparable. Furthermore, we can with a good degree of confidence understand and predict the consequences of things such as ozone loss or CO_2 increase, but we have no clue about the consequences of loss of biodiversity beyond understanding that at some level we will induce the collapse of the biosphere. *We are gambling and the odds are high against us.*

Tri-atomic oxygen

In the 1980s a British research group monitoring the atmosphere over Antarctica reported the shocking result that the level of ozone dropped dramatically every October creating what became known as the "ozone hole." It is not really a hole but rather a rapid decrease in the ozone concentration, which happens over a large area every year around the month of October. What has been observed since its discovery is that the minimum quantity of ozone in the hole has decreased steadily since about 1975, such that by 1995 it was half its value before 1970.

The delicate stratospheric ozone layer has protected life on Earth's surface from damaging solar UV radiation for more than 500 million years.

As we saw in Chapter 4, ozone is produced naturally as solar ultraviolet radiation splits molecular oxygen (O_2) into its two component oxygen atoms, and these can then combine with another oxygen molecule to form tri-atomic ozone (O_3). As ozone absorbs further ultraviolet radiation it splits again into an oxygen molecule and an oxygen atom. These reactions are in equilibrium and maintain a constant average quantity of ozone, as long as the rate of ozone creation equals the rate of destruction. The introduction of substances which can disrupt this balance will affect the final equilibrium quantity of ozone. The complex chemistry of ozone destruction, due to anthropogenic chemicals harmful to this layer, was worked out in the early 1970s by Mario Molina (1943–), Sherwood Rowland (1927–), and Paul Crutzen (1933–). They shared the 1995 Nobel Prize in Chemistry for this work, on a problem of great concern to humanity.

Ozone can be destroyed by bromine which reaches the stratosphere as a byproduct of the release of halons (used in fire extinguishers) and methyl bromide (used as an agricultural fumigant). It can also be destroyed by chlorine, a byproduct of the release of chlorofluorocarbons (CFCs) used in aerosols and as refrigerants. The CFCs were selected for these uses because they are quite unreactive, being non-toxic and non-flammable, something quite convenient if you are going to spray things into your hair or your armpit. However, this property also means that once released into the atmosphere they remain there for long times – one or two hundred years. As these compounds slowly diffuse upward and reach the stratosphere they are broken down by solar ultraviolet radiation, releasing their component bromine and chlorine atoms, which then destroy ozone. This ozone *depletion* is a process which has now been observed over all geographic latitudes, a steady decrease by about 3 percent per year of the average global ozone concentration, clearly discernible over naturally occurring fluctuations. As we have seen, ozone destruction needs ultraviolet sunlight. Since the poles are dark for half of the year, ozone destruction there coincides with the appearance of sunlight, and then proceeds rapidly, creating the hole.

The discovery of the ozone problem led to several national laws and international treaties to limit and phase out the use of compounds which destroy ozone. This did not happen easily. An intense debate, a "Spray-can War," raged in the early 1970s, confronting scientists, industry representatives, and federal agencies in battles fought in the media, in congress, and stridently in otherwise quiet hallways. Uncertainties in the results were often interpreted as meaning that the science was wrong. This might sometimes be the case but in most instances uncertainties are a consequence of the difficult measurements needed and the complexity of the systems one is

"DID HE SAY SOMEDAY WE'D BE SORRY, OR SUNDAY WE'D BE SORRY?"

© Nick Downes

studying. Still, even if in error, I would prefer to err on the safe side, and take the necessary steps to avoid potential problems, knowing that if there were errors these would soon be corrected by further research.

A measured decline of the concentration of these harmful chemicals in the atmosphere has been observed since 1995, a result of the curtailment of their use, showing that where there is a will there is a way. However, because of their long lifetime in the stratosphere, it is expected that ozone depletion will continue far into the first century of this new millennium. It is sobering to reflect that this problem, which almost led to a global catastrophe, resulted from our innocent use of what were thought to be safe products, and also that the discovery that we were in deep trouble was the result of atmospheric research unrelated to the problem. There is a lesson to be learned here.

Our future

All of the above calamities are slow and at first imperceptible. They are the result of very complex processes which are very difficult to disentangle from other possible natural effects. The problems we face are also of such a nature that, by the time the damage is convincingly documented, it might be irreversible, or, as in the case of ozone, difficult to repair. It is as if our planet

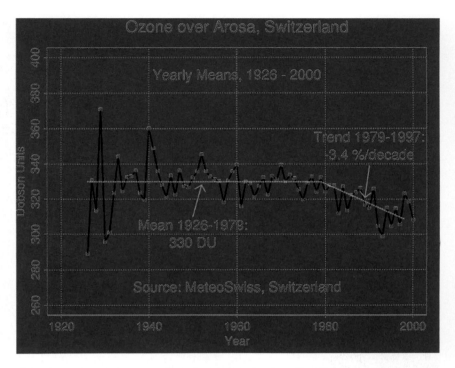

The Dobson spectrophotometer is an instrument developed in 1924 to measure ozone, and its atmospheric concentration is still measured in units as defined by this instrument. The ozone measurements form Arosa, in Switzerland, are the longest available. This graph shows that from 1926 until 1979 the *average* level remained constant, and then started to drop at a rate of 3.4 percent/decade, documenting ozone depletion in the northern mid-latitudes. Ozone depletion is caused by man-made chemicals such as chlorofluorocarbons (CFCs) – compounds consisting of chlorine, fluorine, and carbon – which upon reaching the ozone layer are broken down into their component atoms, allowing the chlorine to react with ozone to yield chlorine monoxide and molecular oxygen. It is estimated that one chlorine atom can destroy over 100 000 ozone molecules before finally being removed from the stratosphere. This is why ozone destruction will continue for many years, even if production of CFCs is completely halted. (MeteoSwiss, Switzerland)

had got some form of cancer. We are easily aware of changes in our neighborhood and you would be very unhappy, and protest vehemently, if someone dumped toxic wastes nearby. If instead, these wastes were diluted and dumped over the entire globe, you would not perceive them and would have no reason to be worried and protest – until it might be too late. And this is what has been happening. The awareness of the threat is provided indirectly by surrogate eyes placed by scientists all over the Earth and in space, and by the frogs. These eyes, contrary to our own, are objective and see the

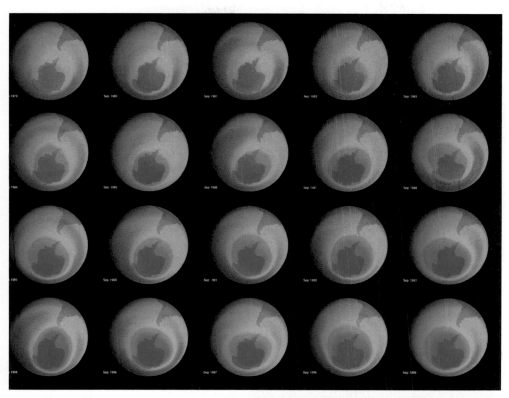

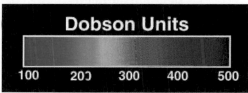

NASA and NOAA instruments have been measuring Antarctic ozone levels since the early 1970s. Large regions of depleted ozone, the "ozone hole," began to develop each year over Antarctica in the early 1980s. The hole becomes deepest between late August and early October of every year. This picture shows the average hole as it appeared in September (close to ozone hole peak) between 1979 to 1999 (from top left to bottom right). Higher levels of ozone are shown in red and yellow and minimum levels in blue as shown by the scale. The continent of Antarctica and the southern part of South America are shaded. The trend of increasing hole depth and size will continue for many years before the situation is remedied by the limitation of the emission of gases harmful to ozone. (NASA GSFC Scientific Visualization Studio)

The influence that humans have had on the Earth can be seen from space on a global scale. Here we see the Earth viewed at night. The light from countless villages, towns, and cities and from rural fires (imaged over several months) light up the night. The major urban population centers are prominent sources of light. (NASA GSFC Scientific Visualization Studio)

planet from a global perspective. It is easy to believe what they see and, as shown in the figures of this chapter, what they see is not pretty.

If all this were not enough, the possibility of committing collective suicide in a nuclear holocaust is ever-present, possibly triggered by a future war fought over dwindling resources. It would be the greatest irony if we ended our existence using the same thermonuclear reactions which for billions of years have provided life-giving energy to Earth. If you think all this paints a somber picture, you are right: *the future looks grim* and is coming inexorably toward us. While some societies are still busy spending their time and energy fighting centuries-old grudges, and politicians argue about all sorts of lesser problems, the world as we know it is coming to an end much sooner than you think. And you know something? I used to be an optimist. Even so, last year I planted nine trees in my garden.

We have been smart enough to understand the basic natural laws which govern physical processes, and have applied this knowledge to develop powerful technologies that for the first time in the history of our planet have artificially altered our global environment. We have also been smart enough to develop sophisticated tools such as computers with intricate software, and space-based remote-sensing instruments, which have allowed us, just in time, to see even more clearly which way we are headed. The really big question then is: are we smart enough to do something about it? You have heard it said that "a little knowledge can be a dangerous thing." In the same vein, I would say that for the longer view a little intelligence can also be dangerous, and I venture to say that this is all we have got, a little intelligence. *We find ourselves in front of the "not-sufficiently-intelligent" hurdle.* It's a pity, isn't it? Of all species on Earth, we are the only one to have sufficient intelligence to at least understand this. A chimp has about one-quarter of our brain capacity but does not seem to achieve even one-thousandth of what we can. Imagine animals with four or even ten times our brain capacity. We cannot imagine what they could. It is above our heads. I believe that this is the tragedy of our existence: enough intelligence to dominate and affect the biosphere, but not enough to have the foresight to avoid its eventual destruction. It is scary to think that our activities, in our one-second appearance in the third hour of the five-hour movie of the story of our planet, have managed to endanger a significant part of the biosphere. This really is just a cameo appearance, but we have certainly stolen the show!

Intelligence is an ambiguous notion, and intelligent beings have spent countless hours trying to define this quality. Maybe the difficulty is due to intelligence not being able to define itself. The Swiss psychologist Jean Piaget (1896–1980) defined intelligence as that which we use when we don't know what to do. Well, we need a lot of it now! We often link intelligence to "goodness," admiring people for their intelligence, although usually this relates to some skill, rather than to what Piaget meant. We are proud and impressed by the great achievements of our species and associate these with intelligence, but I think that intelligence alone is not enough.

To survive we must drastically change our current thinking about what is "good" for humanity, and we must do this unselfishly. We shall have to rely on more than just our treacherous intelligence. I mean a quality which to me is equally important, and equally difficult, to define: we need to be *wise*. It is curious that we ascribe this quality to mythological figures of the past, and to those of our science fiction future. We really need it now.

Just as there have been extinctions in the past that have led to the appearance of new forms of life, it can be argued that if we and many other species perish, new creatures will evolve, we hope more adapted to survive by their intelligence than we are. This assumes that evolution leads to intelligent conscious beings, which as we have seen is not built in or implied by evolution. The only difference would be that this sixth mass extinction would have been caused by our own hands. Still, it is one thing to commit suicide and quite another to commit murder. For better or worse, like it or not, we are the dominant species on this planet, and all life forms, but especially the more complex organisms, depend in great measure on our actions. In this sense, we carry a moral responsibility to think of their well-being, as well as our own, things which are in any case not independent.

So, you may ask, what are we to do? Well, coming up with a solution to all these difficult problems is clearly not easy, since they are probably the most complex ones we ever faced. Although scientists have played a prominent role in pointing out these problems and in suggesting possible solutions, it is not just for science to find solutions to these problems with wide-ranging consequences affecting everyone. Ethical, social, economic, and political considerations also enter the equation of our future. A few steps can and have been taken to alleviate some problems such as the limitation by international agreement of the production of chemicals that attack ozone. This has led to a slowing of ozone destruction, but hardly an end of the problem. About 75 percent of the carbon introduced into the atmosphere from the use of fossil fuels is produced by burning coal to produce electricity. Alternative energy sources to reduce our reliance on coal are known, and it would not take a genius to design toilets that will at least cut in half the 5 billion gallons of water we use daily to flush them. Also, in principle, we know how *not* to make babies. However, as I see it, we are dealing with all this extremely casually and timidly: a conference here, a watered-down regulation there, but mostly business-as-usual in these very unusual times. If you somehow knew that your child risked being hit by a truck tomorrow morning, then you would surely act immediately and with a great sense of urgency. The problems we face seem much less immediate and will affect our children in a more general way, but collectively they are even more dreadful. What we must do is mark them with the highest priority on our global "to do" list.

Time will tell, of course, although many of us will not be around in 2050. We can be bystanders, looking at our own fate as it unfolds like a horror movie, not understanding what drives us blindly down this dangerous path. We can deny it all and believe that we will not be affected, the "we" here being the minority who live in the so-called "developed" countries. Until recently this may have been a viable position as we watched on the evening news, while having dinner, reports of famine and disease in faraway places populated by our poor fellow travelers on spaceship Earth. Now, however, the dimension of the problem is rapidly changing as it becomes truly global and it is no use sticking our heads in the sand; we are all in it together.

Lest you take the view that others are to blame, I remind you that although the population of the developed nations, say the so-called G7 nations (Britain, Canada, France, Germany, Italy, Japan, and the United

States) represents only about 10 percent of the total world population, they consume almost half the fossil fuels and a very large fraction of all commodities. They are also the ones selling cars to other nations which would do better looking for alternative modes of transportation. As long as we accept 10000 new souls onto this planet every hour, *nothing else will work.* Whether we like it or not, whether it is in accord with our most cherished beliefs or not, we need to reduce our numbers and drastically change our ways to have any chance of long-term survival.

We don't seem to have the social and political will to redirect our societies along a path that will avoid a catastrophe; it is not even clear if anyone really knows where this path is. What is clear is that the current path leads to global disaster and that we are running out of time. I suspect that nothing short of what most would call an extreme social revolution will help to fix what can best be described as a *humongous screw-up.*

This planet has large resources but they are limited. It also has a limited capacity to absorb the staggering abuse to which we subject it. We *are* running out of important resources and doing irreparable damage to our environment such that it will become difficult to survive. We are living on the edge, we have pushed our planet to the limits, and we are on the brink of disaster (a word whose etymology appropriately comes from "losing the stars" – of catastrophic consequence to navigators). The problem is that the time scales for natural relaxation of our complex systems are very long compared with human lifetimes, and damage done today can and will have unforeseen consequences in the long-term future. It is also quite clear to anyone thinking about these issues that most of the 9 billion inhabitants expected on our planet in the year 2050 cannot have a lifestyle like the current one in the developed countries. It is not even clear that they will have a lifestyle at all. So the choice is simple: do nothing and suffer the painful consequences, or try to prepare for this very different future which is only 50 years ahead. *We should give it a real try, lest we become the laughingstock of the Galaxy* (assuming that there is someone out there to laugh).

Chapter 9

Epilogue

We apologize for the inconvenience[1]

I hope you have enjoyed this rather short story about a very long story, although let me say that I did not write the previous chapter for you to enjoy. You might even wonder what the last chapter has to do with the rest of the book, but by now you should have noticed that, from the very beginning of this tale, it is the environment that determines the course of events. Throughout the history of our planet, life and the environment have been entwined in a ritual dance, sometimes to fast music with many turns and pirouettes, and at other times, in a tender embrace to slow music. This dance will continue until the music of the spheres stops.

Helium and ice

If our Universe had not been just right, none of this would have happened and I would not now be writing this nor you reading it. This is the great mystery of our existence. The insides of either a star or a dark cosmic cloud, or the surface of a planet at a particular distance from its star, are environments with specific resources and definite limitations, which establish what is possible. The environment of our biosphere is no different; it determines what can happen, and if it changes as it has in the past, or if we change it in the future, then life will be subjected to the consequences, and all might end in tears – our tears.

If I have succeeded, then you are now in a position to understand better your place in the Universe, although you will not have the answer to the question of Life, the Universe, and Everything – certainly it is not 42. For this question a lot of work lies ahead, since, in cosmic terms, we have barely left the starting blocks. Nevertheless, no matter how much we might progress in

---

[1] Douglas Adams, *So Long and Thanks for all the Fish*, Six Stories, p. 611.

our scientific understanding of the natural world, we shall not find meaning by only studying our surroundings. We shall also need to look within ourselves, in a spiritual way, to understand what *we* are all about.

Our intimate connection to the Universe is suggested by the simple fact that both we and it are made of mostly hydrogen. Alternatively, you can explore everything in much more detail, as we have done in this book, and understand in part the complex chain of cosmic events that have led to us. "Primitive" people did not doubt this intimate connection, although they did not understand it. We, who have the tools to understand, seem to have forgotten.

The oxygen and nitrogen we breathe, the aluminum and other metals in our airplanes, the gold or platinum in our rings, and the carbon and other atoms in our bodies were produced in stellar processes. Without stars neither we nor our world would exist, and without the steady source of energy from our Sun, life on our planet could not have developed, nor would it have endured. Many planetary events, some sudden and accidental, others the consequences of the steady evolution of the Earth's surface and its atmosphere, have shaped what we see today. The evolution of life is a cosmic phenomenon, aided by geophysical events which have occurred during the incredibly long history of our planet. Some are as accidental as having a large Moon (moving in the right direction) to stabilize the climate, or our planet being of sufficient mass to retain an atmosphere. Others are the consequences of fundamental physical laws which make ice float and the neutronless isotope of helium unstable. Life can only arise after the elements of life are produced and made available via the cosmic history of stars. The energy of ancient stellar explosions, stored in long-lived radioactive elements, and driving complex geophysical cycles, determines in part the evolutionary process. We have seen how the development of intelligent life is fraught with uncertainties, so much so that the existence of intelligent life on other planets is hardly an obvious matter. On a more fundamental level, had the Universe not expanded from its initial state of high temperature and density, no galaxies, no stars, and no planets would have formed and life would not have emerged.

A new age

At the vernal equinox (March 21), when night and day are of equal length, we see the nearest star, our Sun, against one of the twelve constellations of the zodiac. Because of the precession of Earth's rotation axis, Polaris is only temporarily the north pole star and the vernal equinox travels once around the zodiac in 26 000 years. For the past couple of thousand years it has been in the constellation Pisces. It will enter Aquarius soon, or has already done so, depending on how the borders of the constellations are defined on the celestial sphere. So we are entering the age of Aquarius, an arbitrary and meaningless concept, as is all of astrology, although it does have an attractive ring to it. Nevertheless, we *are* entering a new age, and we can call it Aquarius if we wish. It is an age of conflict between humans and nature, an age in which, through our technology, ultimately based on science, we have gained sufficient power to rapidly upset a billion-year-old balance. In this new age we will run out of some important non-renewable resources such as coal, oil, and natural gas. These formed over many millions of years and we will have consumed them in a few hundred. Once gone, they will be gone for good, like extinct species. Other resources are renewable, but increased consumption resulting from population growth will not allow them to recover. We are witnessing the reduction of available fresh water, forests, fisheries, and many other important resources. *The well will run dry.*

Some 2.5 million years ago our ancestors first chipped flakes from stones to fashion tools. Over most of this stone age (the paleolithic) these tools did not change much. Starting about 100 000 years ago, the record begins to show many innovations including the use of fire by the Neanderthals, the start of the controlled use of combustion, leading to the first use of metals 5000 years ago. The rest is history, culminating with the technological explosion of the last century which gave us unprecedented power. Blinded by this power, we have thought that we could divorce ourselves from nature only to realize that we will be left with the short end of the stick. We are a part of nature and need to go back to her with *love and respect*, the formula for any good relationship. We cannot ignore all we have learned. We cannot ignore our discoveries about our world and hide behind the presumed safety of dogma. No inquisition can change the way the world is. Yes, back to nature, but not in the sense of repudiating science and technology. We cannot put the genie back in the bottle. Quite the contrary, we need science more than ever, to gain an even deeper understanding of nature and our relationship with her. We must use technology wisely to see if it might be possible to extricate ourselves from this "humongous screw- up."

Human life appeared on Earth's biosphere after a long evolutionary process operating under a fragile equilibrium. As you have seen, we are the result of an amazing process, but surely not the end result, and definitely not a necessary one. The history of life on Earth teaches us that we are here because of a lucky break, in fact several of them. Let us not press our luck lest nature decides she has had enough of us and the age of Aquarius turns into the last dark age.

We should not waste that incredible gift which distinguishes us from, but makes us in no sense better than, other animals on Earth: *our minds*. Thanks to our minds, we may understand what life is all about, and may somehow make a difference. We will not overcome the "not-sufficient-intelligence" barrier by evolving into a new, more intelligent, species. Because we have filled the global ecosystem, we have reached the end of biological evolution which needs isolated small groups to proceed. And, even if the above development were possible, it would not necessarily move in the direction of more intelligence. I discard the notion of artificial evolution by genome manipulation (suggested by some) as not wise, for the simple reason that we should not act as if we were the gods. Our only hope to jump the hurdle is through cultural evolution, by establishing new values and defining a new and very different vision for the future. In this we will be aided by better intelligence provided by the computers and software of the future which might help us to see farther and deeper, allowing us to become wiser. In the long run this might even allow us to escape the fate of all other complex species which have inhabited Earth – extinction. I worry that we might not have sufficient time for this, but it offers us a glimmer of hope. If we do not survive, then all the sweat, blood, and tears expended throughout history will have been for nothing, and all our thoughts about Life, the Universe, and Everything, utterly meaningless. A depressing thought.

Children of the stars

Five hundred years ago we had placed ourselves at the center of the stage of a theater built especially for us. It was the only show in town, written just for us, running forever – immutable. Copernicus began our demotion and today we understand that we inhabit an insignificant planet which is literally in the middle of nowhere. Geophysics took away the sense of permanence and security by telling us that, although we seldom perceive it, the stage is changing all the time. We have learned that our participation has been merely a small part in a very long and varied show. Evolution added to the insult by taking away our last pretense of being special, pointing out that we were only extras chosen at random to appear for the blink of an eye, and not really for any particular talent we might presume to have. Our only consolation is that, because we are privileged to have a large brain, we can follow the great show, although it may turn out to be a tragedy.

We are *children* because we have just arrived to start our journey, and like children we are disorderly and do things which are not in our best interest. It may be more appropriate to think of ourselves as *orphans*, having been left

We are children of the stars. This breathtaking image shows a cosmic cauldron, a cluster of brilliant massive stars (lower right) that is located at the edge of the Tarantula nebula of the Large Magellanic cloud, 160 000 light-years away. The multiple sheets and filaments seen at the upper left were produced by several supernova explosions which sent shock waves that compressed the surrounding gas. Small regions of dense gas clouds have formed near the center where new stars are being born. Planetary systems might form around some of these stars and, if the conditions are just right, life might appear on one of them. (AURA/STScI/NASA)

behind on this minuscule island we call Earth to fend for ourselves. The perils are everywhere and there are no adults to warn and guide us to what is best for us, although we have invented gods to play this role. The future is quite uncertain: unknown dangers lurk ahead, deep waters, treacherous swamps, and dangerous creatures, maybe the greatest threat being our own actions whose consequences we have only recently begun to recognize. After this journey through space and time, the thought that we are *Children of the Stars* becomes evident. Our subconscious yearning to explore the distant realms of the Universe that we are a part of does not surprise us. In some sense our genes make us build telescopes, microscopes, and particle accelerators to study our origin, and eventually to find out if we are all alone in this incredibly vast Universe. They also made us go to the Moon, and will make us strive to reach our parents: the stars.

We have been fruitful and have multiplied, have filled the Earth and subdued it. We have dominion over the birds of the air and over the living things that move upon the Earth. It remains to be seen if this is good. Please take care of our planet; it is the only one we've got, and for all we know it might be the only one with life such as ours in a very large region of the Universe. You are here for not even 100 years, but others will follow. Go out and contemplate the stars tonight; you will see them in a very different light.

Further reading

I hope that you are now wondering what to read next, maybe a book which delves into the topics covered in this book in more detail, but is not difficult to read. The choice is abundant, being such a fascinating subject, where new discoveries are reported every year, and I can only propose that you read what I consider to be the most appropriate of those which I have read. No doubt there are many others which I omit because I am not familiar with them.

If you want to know about the whole shebang then read the book by Timothy Ferris *The Whole Shebang* (what else?) (Simon & Schuster, 1998). The life and evolution of stars and cosmic clouds are laid out in great detail in two nicely illustrated books by James B. Kaler: *Stars* (W.H. Freeman, 1998) and *Cosmic Clouds − Birth Death and Recycling in the Galaxy* (W.H. Freeman, 1997). These two are more technical than the first but not difficult.

For a very readable and enjoyable account of our changing view of the Universe and an introduction to modern cosmology, I recommend *Blind Watchers of the Sky* by Rocky Kolb (Addison Wesley, 1996) or *The Dancing Universe* by Marcelo Gleiser (Dutton, 1997). A wonderful book about the history of our changing vision of the Universe is Arthur Koestler's *The Sleepwalkers* (Penguin, 1964). A book that is much more than its title implies is *Venus Revealed* by Harry Grinspoon (Addison Wesley, 1997), an enjoyable book that covers many of the topics discussed here. A beautiful book about our solar system is *The New Solar System* edited by J. Kelly Beatty, Carolyn Collins Petersen and Andrew Chaikin (Cambridge University Press, 1999). A fast-paced story about the origin and evolution of the solar system is given by Stuart Ross Taylor in *Destiny or Chance − our Solar System and its Place in the Cosmos* (Cambridge University Press, 1998). You can find a more in-depth discussion of many of the topics covered here in *Rare Earth* by Peter D. Ward and Donald Brownlee (Copernicus, 2000).

If you are intrigued by the chapters on life, I can recommend two excellent books which will satisfy your curiosity. I enjoyed and learned a lot from Christian de Duve's *Vital Dust – The Origin and Evolution of Life on Earth* (Basic Books, 1995). This is a book to be read with time and is a rewarding experience. Or you might try Paul Davies's *The Fifth Miracle – The Search for the Origin and Meaning of Life* (Simon and Schuster, 1999), shorter and more fun to read, and a good one to try first. They both treat their topics with significant depth and erudition. *Microcosmos – Four Billion Years of Microbial Evolution* (University of California Press, 1997), written by Lynn Margulis and Dorion Sagan, is also engrossing. Nicely illustrated and based on an exhibition is *A Walk Through Time – From Stardust to Us* by S. Liebes, E. Sahtouris, and B Swimme (Wiley, 1998), concentrating mainly on the development of life on Earth.

The story of the K–T extinction and the discovery of its causes is well told by its principal protagonist in Walter Alvarez's *T. Rex and the Crater of Doom* (Princeton University Press, 1997). I highly recommend this one to you.

If you wish to read more about Mother Earth try J. MacDougall's *A Short History of Planet Earth* (Wiley, 1998), a book which provides a concise introduction to the development of our planet and life.

If SETI is what you want to learn more about you can read Seth Shostak's *Sharing the Universe* (Berkeley Hills, 1998), a witty, easy to read, discussion of the topic.

The origin of modern humans, a topic not covered in this book, is the subject of an excellent work entitled, not surprisingly, *The Origin of Modern Humans* by Roger Lewin (W.H. Freeman, 1998).

Should you be concerned about our future, and also find yourself a bit confused by the different things you hear in respect to our environment, then read the excellent book *Betrayal of Science and Reason* (Island Press, 1996) by two prominent researchers who have long been at the forefront of the environmental movement: Paul and Anne Ehrlich. It will open your eyes. A short, easy-to-read book dealing with our current predicament and how it came about is *Dominion* by Niles Eldrege (University of California Press, 1995). For a broader discussion including many interesting details read *The Spirit in the Gene* by Reg Morrison (Cornell University Press, 1999). A book that goes into more depth on some of the topics covered here is *Our Cosmic Origins* by Armand Delsemme (Cambridge University Press, 1998). Finally, *The Diversity of Life* by Edward O. Wilson (Norton, 1999) is a beautiful book. Just read it.

Two outstanding recent books cover different aspects of this story and would complement what you might have learned here. One is the very readable *Stardust* by John Gribbin (Yale University Press, 2000) and the other one, at a somewhat more technical level, is *Cosmic Evolution* by Eric Chaisson (Harvard University Press, 2001).

The Ultimate Hitchiker's Guide is by Douglas Adams (Wings Books, 1996).

If you want to keep up-to-date on these topics you could buy or, better still, subscribe to monthly magazines such as *Sky and Telescope* or *Astronomy*.

Appendix A

The story of the Sibylline books

The story of the Sibylline books[1] concerns an ancient city – it doesn't matter where it was or what it was called – it was a thriving, prosperous city set in the middle of a large plain. One summer, while the people of the city were busy thriving and prospering away, a strange old beggar woman arrived at the gates carrying twelve large books, which she offered to sell them. She said that the books contained all the knowledge and all the wisdom of the world, and that she would let the city have all twelve of them in return for a single sack of gold.

The people of the city thought this was a very funny idea. They said she obviously had no conception of the value of gold and that probably the best thing was for her to go away again. This she agreed to do, but first, she said, she was going to destroy half of the books in front of them. She built a small bonfire, burnt six of the books of all knowledge and all wisdom in the sight of the people of the city and then went on her way. Winter came and went, a hard winter, but the city just about managed to flourish through it and then, the following summer the old woman was back.

"Oh, you again," said the people of the city. "How's the knowledge and wisdom going?"

"Six books," she said, "just six left. Half of all the knowledge and wisdom in the world. Once again I am offering to sell them to you."

"Oh, yes?" sniggered the people of the city.

"Only the price has changed."

"Not surprised."

"Two sacks of gold."

"What?"

[1] The Sibyls are legendary seeresses of antiquity. It is not clear how many there were nor where they came from. This story is taken from D. Adams and M. Carwardine *Last Chance To See*, 2nd edn, Pan Books Ltd and William Heinemann Ltd, London, 1991, pp. 196–9 reprinted with the permission of Douglas Adams.

"Two sacks of gold for the six remaining books of knowledge and wisdom. Take it or leave it."

"It seems to us," said the people of the city, "that you can't be very wise or knowledgeable yourself or you would realize that you can't just go around quadrupling an already outrageous price in a buyer's market. If that's the sort of knowledge and wisdom you're peddling then, frankly, you can keep it at any price."

"Do you want them or not?"

"No."

"Very well. I will trouble you for a little firewood."

She built another bonfire, and burnt three of the remaining books in front of them and then set off back across the plain.

That night one or two curious people from the city sneaked out and sifted through the embers to see if they could salvage the odd page or two, but the fire had burnt very thoroughly and the old woman had raked the ashes. There was nothing.

Another hard winter took its toll on the city and they had a little trouble with famine and disease, but trade was good and they were in reasonably good shape again by the following summer when, once again, the old woman appeared.

"You're early this year," they said to her.

"Less to carry," she explained, showing them the three books she was still carrying. "A quarter of all the knowledge and wisdom in the world. Do you want it?"

"What's the price?"

"Four sacks of gold."

"You're completely mad, old woman. Apart from anything else our economy's going through a bit of a sticky patch at the moment. Sacks of gold are completely out of the question."

"Firewood, please."

"Now wait a minute," said the people of the city, "this isn't doing anybody any good. We've been thinking about all this and we've put together a small committee to have a look at these books of yours. Let us evaluate them for a few months, see if they're worth anything to us, and when you come back next year perhaps we can put in some kind of a reasonable offer. We are not talking sacks of gold here, though."

The old woman shook her head. "No," she said. "Bring me the firewood."

"It will cost you."

"No matter," said the woman, with a shrug. "The books will burn quite well by themselves."

So saying she set about shredding two of the books into pieces which then burnt easily. She set off across the plain and left the people of the city to face another year.

She was back in the late spring.

"Just the one left," she said, putting it down on the ground in front of her. "So I was able to bring my own firewood."

"How much?" said the people of the city.

"Sixteen sacks of gold."

"We'd only budgeted for eight."

"Take it or leave it."

"Wait here."

The people of the city went off into a huddle and returned half an hour later.

"Sixteen sacks is all we've got left," they pleaded. "Times are hard. You must leave us with something."

The old woman just hummed to herself as she started to pile the kindling together.

"All right!" they cried at last, opened up the gates of the city and let out two oxcarts, each laden with eight sacks of gold, "but it had better be good."

"Thank you," said the old woman, "it is. And you should have seen the rest of it."

She led the two oxcarts away across the plain with her, and left the people of the city to survive as best they could with the one remaining twelfth of all the knowledge and wisdom that had been in the world.

Appendix B

Some numbers

The Earth

Mass	5.974×10^{24} kilograms ($\sim$ 80 Moons)
Equatorial radius	6378.256 km $\sim$ 3,956 miles
Period of orbit	365.256 days (one year)
Speed around Sun	$\sim$108 000 kilometers per hour ($\sim$67 000 miles per hour)
Solar irradiation	$\sim$1368 watts per square meter
Composition of atmosphere	$\sim$80% nitrogen, 20% oxygen
Main composition	$\sim$35% iron, 30% oxygen, 15% silicon, 13% magnesium
Density	$\sim$5.52 times that of water
Escape velocity	11.2 km/s ($\sim$25 000 miles/hour)

The Moon

Distance from Earth	$\sim$384 000 km ($\sim$240 000 miles or 1.25 light-seconds)
Mass	$\sim$7.4 $\times$ 10^{22} kilograms, $\sim\frac{1}{80}$ Earth
Radius	$\sim$1738 km ($\sim$1080 miles)
Time between full phase	$\sim$29.5 days
Speed around Earth	$\sim$3400 km/h ($\sim$2100 miles/h)
Density	$\sim$3.34 times that of water

The Sun

Distance from Earth	$\sim$ 150 000 000 km ($\sim$93 000 000 miles or 8 light-minutes)

Mass	$\sim 2 \times 10^{30}$ kilograms ($\sim 335\,000$ Earths)
Radius	$\sim 700\,000$ km (~ 110 Earths)
Luminosity	$\sim 4 \times 10^{26}$ watts
Surface temperature	~ 6000 K
Central temperature	$\sim 15{,}000{,}000$ C
Composition	$\sim 92\%$ H , 8% He and traces of O, C, Ne, Mg, N, Si, Fe, S
Density (average)	~ 1.41 times that of water

Notes

The quantity 4×10^{26} equals $400\,000\,000\,000\,000\,000\,000\,000\,000$.
and more generally $a \times 10^{b}$ equals an a followed by b zeros.
One kilogram is a mass that weighs 2.2 pounds.
The density of water is 1000 kilograms per cubic meter.

Appendix C

World scientists' warning to humanity

Some 1700 of the world's leading scientists, including the majority of Nobel laureates in the sciences, issued this appeal in *November 1992*. The World Scientists' Warning to Humanity was written and spearheaded by the late Henry Kendall, former chair of the *Union of Concerned Scientists* board of directors.

Introduction

Human beings and the natural world are on a collision course. Human activities inflict harsh and often irreversible damage on the environment and on critical resources. If not checked, many of our current practices put at serious risk the future that we wish for human society and the plant and animal kingdoms, and may so alter the living world that it will be unable to sustain life in the manner that we know. Fundamental changes are urgent if we are to avoid the collision our present course will bring about.

The environment

The environment is suffering critical stress:

The Atmosphere

Stratospheric ozone depletion threatens us with enhanced ultraviolet radiation at the earth's surface, which can be damaging or lethal to many life forms. Air pollution near ground level, and acid precipitation, are already causing widespread injury to humans, forests, and crops.

Water Resources

Heedless exploitation of depletable ground water supplies endangers food production and other essential human systems. Heavy demands on the

world's surface waters have resulted in serious shortages in some 80
countries, containing 40 percent of the world's population. Pollution of
rivers, lakes, and ground water further limits the supply.

Oceans

Destructive pressure on the oceans is severe, particularly in the coastal
regions which produce most of the world's food fish. The total marine catch
is now at or above the estimated maximum sustainable yield. Some
fisheries have already shown signs of collapse. Rivers carrying heavy
burdens of eroded soil into the seas also carry industrial, municipal,
agricultural, and livestock waste – some of it toxic.

Soil

Loss of soil productivity, which is causing extensive land abandonment, is a
widespread by-product of current practices in agriculture and animal
husbandry. Since 1945, 11 percent of the earth's vegetated surface has been
degraded – an area larger than India and China combined – and per capita
food production in many parts of the world is decreasing

Forests

Tropical rain forests, as well as tropical and temperate dry forests, are being
destroyed rapidly. At present rates, some critical forest types will be gone in
a few years, and most of the tropical rain forest will be gone before the end
of the next century. With them will go large numbers of plant and animal
species.

Living Species

The irreversible loss of species, which by 2100 may reach one-third of all
species now living, is especially serious. We are losing the potential they
hold for providing medicinal and other benefits, and the contribution that
genetic diversity of life forms gives to the robustness of the world's
biological systems and to the astonishing beauty of the earth itself. Much
of this damage is irreversible on a scale of centuries, or permanent. Other
processes appear to pose additional threats. Increasing levels of gases in the
atmosphere from human activities, including carbon dioxide released from
fossil fuel burning and from deforestation, may alter climate on a global
scale. Predictions of global warming are still uncertain – with projected
effects ranging from tolerable to very severe – but the potential risks are
very great.

Our massive tampering with the world's interdependent web of life –
coupled with the environmental damage inflicted by deforestation, species

loss, and climate change – could trigger widespread adverse effects, including unpredictable collapses of critical biological systems whose interactions and dynamics we only imperfectly understand. Uncertainty over the extent of these effects cannot excuse complacency or delay in facing the threats.

Population

The earth is finite. Its ability to absorb wastes and destructive effluent is finite. Its ability to provide food and energy is finite. Its ability to provide for growing numbers of people is finite. And we are fast approaching many of the earth's limits. Current economic practices which damage the environment, in both developed and underdeveloped nations, cannot be continued without the risk that vital global systems will be damaged beyond repair.

Pressures resulting from unrestrained population growth put demands on the natural world that can overwhelm any efforts to achieve a sustainable future. If we are to halt the destruction of our environment, we must accept limits to that growth. A World Bank estimate indicates that world population will not stabilize at less than 12.4 billion, while the United Nations concludes that the eventual total could reach 14 billion, a near tripling of today's 5.4 billion. But, even at this moment, one person in five lives in absolute poverty without enough to eat, and one in ten suffers serious malnutrition.

No more than one or a few decades remain before the chance to avert the threats we now confront will be lost and the prospects for humanity immeasurably diminished.

Warning

We the undersigned, senior members of the world's scientific community, hereby warn all humanity of what lies ahead. A great change in our stewardship of the earth and the life on it is required, if vast human misery is to be avoided and our global home on this planet is not to be irretrievably mutilated.

What we must do

Five inextricably linked areas must be addressed simultaneously: We must bring environmentally damaging activities under control to restore and protect the integrity of the earth's systems we depend on.

We must, for example, move away from fossil fuels to more benign, inexhaustible energy sources to cut greenhouse gas emissions and the pollution of our air and water. Priority must be given to the development of energy sources matched to Third World needs – small-scale and relatively easy to implement.

We must halt deforestation, injury to and loss of agricultural land, and the loss of terrestrial and marine plant and animal species.

We must manage resources crucial to human welfare more effectively.

We must give high priority to efficient use of energy, water, and other materials, including expansion of conservation and recycling.

We must stabilize population. This will be possible only if all nations recognize that it requires improved social and economic conditions, and the adoption of effective, voluntary family planning.

We must reduce and eventually eliminate poverty.

We must ensure sexual equality, and guarantee women control over their own reproductive decisions.

Developed nations must act now

The developed nations are the largest polluters in the world today. They must greatly reduce their overconsumption, if we are to reduce pressures on resources and the global environment. The developed nations have the obligation to provide aid and support to developing nations, because only the developed nations have the financial resources and the technical skills for these tasks.

Acting on this recognition is not altruism, but enlightened self-interest: whether industrialized or not, we all have but one lifeboat. No nation can escape from injury when global biological systems are damaged. No nation can escape from conflicts over increasingly scarce resources. In addition, environmental and economic instabilities will cause mass migrations with incalculable consequences for developed and undeveloped nations alike.

Developing nations must realize that environmental damage is one of the gravest threats they face, and that attempts to blunt it will be overwhelmed if their populations go unchecked. The greatest peril is to become trapped in spirals of environmental decline, poverty, and unrest, leading to social, economic, and environmental collapse.

Success in this global endeavor will require a great reduction in violence and war. Resources now devoted to the preparation and conduct of war – amounting to over $1 trillion annually – will be badly needed in the new tasks and should be diverted to the new challenges.

A new ethic is required – a new attitude towards discharging our responsibility for caring for ourselves and for the earth. We must recognize the earth's limited capacity to provide for us. We must recognize its fragility. We must no longer allow it to be ravaged. This ethic must motivate a great movement, convincing reluctant leaders and reluctant governments and reluctant peoples themselves to effect the needed changes.

The scientists issuing this warning hope that our message will reach and affect people everywhere. We need the help of many.

We require the help of the world community of scientists – natural, social, economic, and political.

We require the help of the world's business and industrial leaders.

We require the help of the world's religious leaders.

We require the help of the world's peoples.

We call on all to join us in this task.

Index